R for Programmers

Mastering the Tools

R for Programmers

Mastering the Tools

Dan Zhang

CRC Press
Taylor & Francis Group
Boca Raton London New York

CRC Press is an imprint of the
Taylor & Francis Group, an **Informa** business

Published with arrangement with the original publisher, Beijing Huazhang Graphics and Information Company.

CRC Press
Taylor & Francis Group
6000 Broken Sound Parkway NW, Suite 300
Boca Raton, FL 33487-2742

© 2016 by Taylor & Francis Group, LLC
CRC Press is an imprint of Taylor & Francis Group, an Informa business

No claim to original U.S. Government works

International Standard Book Number-13: 978-1-4987-3681-7 (Paperback)

Visit the Taylor & Francis Web site at
http://www.taylorandfrancis.com

and the CRC Press Web site at
http://www.crcpress.com

Contents

SECTION II R SERVER

SECTION III DATABASE AND BIG DATA

SECTION IV   APPENDIXES

Preface

Why This Book?

As a programmer, I've worked in the area of program development for 10 years. Being a programmer at the very beginning and now an architect, during these 10 years I've experienced many systems and applications. I've developed mobile games as well as programming tools. I've worked in large Web application systems, internal Customer Relationship Management (CRM) of companies, system integration of Service-Oriented Architecture (SOA), and big data tools based on Hadoop. Outsourcing, E-commerce, group purchase, payment, Social Networking Service (SNS), and mobile SNS are all within my working range. In the past I used only Java, then I started learning to use Hypertext Preprocessor (PHP). Just as other programmers, I was exhilarated chasing all kinds of technical innovations. But I was always puzzled by a problem, that is, how to transform the techniques I master into real value. It's as if I had access to a gold mine. With all the advanced techniques I owned I could make mining machines that were stable in performance and outstanding in function, but what I didn't know was how to purify the ore and turn it into gold! I could only stay envious and unwilling every time I saw others getting their gold using my technique.

All of this came to an end when I finally met R. R has opened the gate of treasures from another perspective. It also helped me to start rethinking and replanning my career and finally confirmed my decision to transition to statistics and finance. If the aforementioned problem also occurs in your career, maybe we should get into the world of R and appreciate the special charm of R together in this book. Learning to use R will help us rediscover the value of big data and further raise our personal career value.

As I gradually deepen the exchange with my friends in statistics and finance, I realize that they are also sometimes puzzled and confused about the actual usage of R. For instance, they can achieve an expected outcome with R in a laboratory environment. But when they transplanted their computing to the real environment with complex big data, they often found numerous problems using R. It's as if they had the most advanced purification technique, but their tools for mining and collecting still remained in the Stone Age. Their outdated tools forced them to deal with a mass of problems beyond R, which made them quite overwhelmed. Thus some of them even started to think that R was only a laboratory programming language, or at least it was impossible to apply R in real practice with the state-of-the-art technique. Therefore the extensive application of R would be rather unlikely.

If you are a user of R without any computer background, you may have come across similar problems in your work. Maybe you've spent many sleepless nights trying various ways to find the solutions to these problems. However, we already have mature and effective solutions in the computer industry.

All of the content of this book comes from my summary of my experiences of using R in my work, which means it is rather a real record of my working with R. The content covers the fields of computing, Internet, database, big data, statistics, and finance. This book is a detailed summary of all the solutions of the comprehensive application of R with Java, MySQL, Redis, MongoDB, Cassandra, Hadoop, Hive, HBase, and so forth, which is strong in practice and operability. If you are a beginner user of R, this book may help you appreciate the charm of R in all industries and all fields. If you have used R for a while in a certain industry, this book may show a picture of strong vitality when combining R with other computer languages and help you break through the bottleneck. If you work in technical areas, the case implementation from a holistic view in this book may bring you a new enlightenment, or even help you replan your career and find a new orientation for studying and striving, just like it did for me. If you are a middle or senior manager of a corporation, you will find many technical achievements in this book. You may even directly apply these achievements in a corporate environment and make profits according to the detailed operation records in this book.

Here I should point out that this book is not an introduction or guide, which means there will be no syntax explanation of R in this book. You may have picked the wrong book if you wish to gain basic and introductory knowledge of R. However, if you are familiar with the basics of R but lack a computer language background, this book will tell you what R can do in a real environment and how to implement the applications step by step.

After discussing with many beginners in R from different fields, I find that the greatest problem in learning to use R is how to manage to use its numerous software packages. There are few books and only some pamphlets on the Internet covering this problem. This book covers more than 30 packages of R with my experiences in practice and case analysis. I believe this will help us solve problems using the packages of R.

This book is the first one of the series *A Geek Ideal of R*. Its companion piece, *R for Programmers: Advanced Techniques*, will give an in-depth introduction to the underlying principle of R and how to develop enterprise applications with R.

The environment involved in this book includes two operating systems, Linux Ubuntu® and Windows® 7 and the 2.15.3 version and 3.0.1 version of R, which are specifically identified in each section.

R is constantly advancing and undergoing updates, and it will lead a revolution of data eventually. Interdisciplinary integration is the trend of this development, which is also a great opportunity for us.

Potential Readers of This Book

This book will be helpful to the following people working with R:

- Software engineers with a computer background
- DBA with database background
- Data scientists with a data analysis background
- Scientific researchers with a statistical background
- Students in universities and colleges

How to Read This Book

The content of this book is divided into four sections.

The first section consists of the basics of R (Chapters 1 to 3), which introduces why we should learn to use R, the installation of different versions of R, and the 12 frequently used packages of R. This section will help readers quickly understand the tool packages, time series packages, and performance monitoring packages of R.

The second section discusses the server of R (Chapters 4 and 5), which introduces communication between R and other programming languages and application of R as servers. This section will help readers integrate R with other programming languages and implement the server application of R.

The third section discusses database and big data (Chapters 6 and 7), which introduce communication between R and various databases, as well as R's integration with Hadoop. This section will help readers integrate R with the underlying level of other databases, and implement the processing of big data by R based on Hadoop.

The fourth section comprises the appendixes, which introduce the installation of Java, various databases, and Hadoop. The author expects that readers can accomplish all the cases in this book without other references.

Because this is a reference book, there is no special sequence for reading all the chapters. You can choose the chapters in which you have an interest to start reading. If you are a beginner of R and you wish to master R comprehensively, please follow the chapters in sequence.

Correction and Support

Because the time spent writing this book and the author's knowledge of R are both limited, there will inevitably be some errors and incorrect viewpoints in this book. I sincerely hope that readers will point out and comment on any errors. For this purpose, I have created an online communication website for this book's readers (http://onbook.me) to use to communicate. If you encounter any problems reading this book, please leave notes on this website, and I will try my best to offer a satisfactory solution. All of the source code of this book can be downloaded from the official website of CRC Press (https://www.crcpress.com) or from the online communication website, where I will update the codes in time. This book is printed in black and white, so all the colorful pictures can only be achieved by running the codes of this book. I sincerely hope that you can send your valuable feedback and advice on this book to bsspirit@gmail.com.

About the Translator

Tong Tong, a fan of R language like me, is currently working in an import and export bank in China. He specializes in RMB trade, which makes him an expert with profound knowledge and understanding of the internationalization of the RMB. Tong Tong graduated from China Foreign Affairs University, where he majored in international economics and trade. He also has a background in statistics and finance.

Tong Tong always has the passion to share his practical experience and combine his daily work with the R language. I had the privilege of meeting him in a financial forum and I communicated a lot with him about the use of R language. Despite the fact that Tong Tong does not have a pure computer background, his passion for R language also made him discover a lot of unique ideas in the past. As a self-learner, he did encounter many problems during the practice and he finally found solutions through the book *R for Programmers: Mastering the Tools.*

Hereby, I am honored to invite Tong Tong to help me spread knowledge about R to many readers with diverse mother tongues, and we also hope that it is a chance to share with the world how we use the R language in China.

Once again, thank you, Tong Tong!

Acknowledgments

I wish to thank my teammates Lin Weilin, Lin Weiping, and Deng Yishuo. Thanks to R for bringing us together. I thank He Ruijun, acquiring editor at CRC Press, who helped promote the publication of this book. I also thank the translator, Tong Tong, for the translation work on this book. I give special thanks to my parents and my wife for their support of my work and their care.

This book is dedicated to my dearest family and the fans of R.

BASICS OF R

I

Chapter 1

Basic R Packages

This chapter first presents reasons for learning to use R. It then considers the installation, development tools, and a few commonly used packages of R to help readers gain a quick acquaintance of R and stimulate their interest in learning the language.

1.1 R Is the Most Worthwhile Programming Language to Learn

Question
Why should we learn to use R?

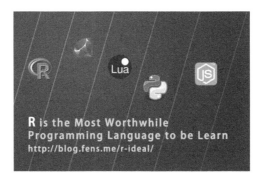

Among the five programming languages—Node, Lua, Python, Ruby, and R—which one will have the best application prospects in 2014 in China?

My choice is R, and I think R will be a star programming language not only in 2014 but also for a long time in the future. This book therefore begins with a discussion of why R is the most worthwhile programming language to learn.

1.1.1 Experience with Java

As a programmer and architect, I had long believed, from the start, that Java would be the programming language to change the world. To some extent this prediction came true; Java has changed the world and performed brilliantly, becoming more powerful and covering more fields. But it turns out that with this growth Java starts to show its limitations, which provides a great opportunity to develop other programming languages.

I've used Java for 12 years, R for 4 years, and Node for 2 years, and as for the question of which programming language had the best application prospect in 2014, my answer would be R.

1.1.2 Why Choose R?

In this section I will elaborate on the reasons to choose R based on the following perspectives.

- Origin of R
- Development of R
- Communities and resources of R
- Philosophy of R
- Users of R
- Syntax of R
- Thinking pattern of R
- Problems to be solved by R
- Shortcomings of R

1.1.2.1 Origin of R

In 1992, Ross Ihaka and Robert Gentlemen, two statisticians from University of Auckland of New Zealand, invented a new programming language to teach elementary statistics courses more conveniently. As both statisticians have R as their first initial, R was adopted as the name of this newly invented programming language.

I have started a cross-border approach to knowledge since learning to use R. Statistics is based on probability, while probability is based on mathematics. R is based on the premise that we are trying to solve practical questions in statistics through programming. The intersection of the knowledge of various subjects will determine our ability to solve the problems. The generic functions of R in statistics make it a distinctive programming language.

1.1.2.2 Development of R

For a long time, R was used in a minority area, initially only by statisticians who wished to replace SAS (Statistical Analysis System). As the concept of big data became more widespread, R was finally discovered by industry. Subsequently, more and more people with engineering backgrounds started to join the circle and made many improvements and upgrades to the computing engine, performance, and a variety of programming packages of R, which led to the rebirth of R reborn as a powerful new language.

The R used today has come much closer to the standard of industrial software. Driven by engineers instead of only statisticians, R has gained a much more rapid growth. As the demand of data analysis continues to grow, R will achieve a faster development and become synonym for free and open data analysis software.

1.1.2.3 Communities and Resources of R

The development of R cannot be attained without the support of the various communities of R. Admittedly, the official R website (http://www.r-project.org/) is too simple and crude as a web-page. Just a small adjustment of the CSS style sheet will give the website a much better appearance, although a simple and unadorned style may appeal to statisticians.

In the official R website, we can download the R software as well as many third-party packages and other supportive software. We can find much information concerning R, such as the developer forum (http://r.789695.n4.nabble.com/), the R-Journal list (http://journal.r-project.org/), package list, R book list, and R user groups. The resources provided by communities of R are enormously abundant, just as for other programming languages.

R is free software, which means that developers can package certain functions in their own packages and publish them on CRAN (http://cran.rstudio.com/). As of October 2015, the total number of R packages published on CRAN is 6910.

Some may doubt that 6910 as the number of packages is relatively small compared to other software. However, all the R packages submitted to CRAN should be reviewed and examined by R group before their publication. The review and examination processes of R packages are very strict, which makes high quality a basic requirement for all the packages published on CRAN. And because many developers find the review process too strict, they opt to publish their packages on RForge (https://r-forge.r-project.org/). Some other R packages are based on Github. I've also published some R packages on Github, https://github.com/bsspirit/chinaWeather.

The following are some main communities and resources of R.

- R official website: http://www.r-project.org/
- Developer forum of R: http://r.789695.n4.nabble.com/
- CRAN: http://cran.rstudio.com/
- RForge: https://r-forge.r-project.org/
- News and blogs of R: http://www.r-bloggers.com/
- Capital of Statistics: http://cos.name/

1.1.2.4 Philosophy of R

Every programming language has its own design concept and philosophy. As for my experience, the philosophy of R is to get down to work.

We do not need to write long codes or design certain models using R. We can achieve a complex statistical model just by a function call and entering some parameters. It is about which model and what parameters to choose, rather than about how to program.

Using R, we will turn a mathematical formula into a statistical model, and we may also consider how to make the result of a classifier more accurate. But we will not think about the time and space complexity in the application of R.

The philosophy of R can transfer your knowledge of mathematics and statistics into computing models. And that is also determined by the origin of R.

1.1.2.5 Users of R

As noted previously, at first R was used only by some statisticians in academia, and then it became widely adopted by scholars in many other fields. The applications of R can cover many fields including statistical analysis, applied mathematics, financial analysis, economic analysis, social sciences, data mining, artificial intelligence, bioinformatics, biopharmaceuticals, global geographical science, data visualization, and so forth.

The big data revolution triggered by the Internet in recent years has led many experts in industry to start to learn and use R. With this expansion of interest, R has gradually met the demand of industrialization and achieved full field development.

The following are some R packages that help promote the development of R in industry.

- RHadoop products of Revolution Analytics, which allows R to call the cluster resources of Hadoop
- RStudio products of RStudio, which gives us some new understanding of editing software.
- RMySQL, ROracle and RJDBC, which create channels for R to visit databases
- rmongodb, rredis, RHive, rhbase, RCassandra, which create channels for R to visit NoSQL
- Rmpi and snow, which make parallel computing of a stand-alone equipment with multicore possible
- Rserve and rwebsocket, which allow R to make data communication among platforms

1.1.2.6 Syntax of R

Similar to Python, R is an object-oriented programming language, but the syntax of R is much freer. The names of many R functions are quite arbitrary, which may be part of the philosophy of R.

A programmer with a foundation in programming languages other than R may possibly feel disconcerted when he or she sees assignment syntax as follows.

```
> a<-c(1,2,3,4)->b
> a
[1] 1 2 3 4
> b
[1] 1 2 3 4
```

But it is so easy to randomly pick out 10 numbers in a N(0,1) normal distribution.

```
> rnorm(10)
 [1] -0.694541401  1.877780959 -0.178608091  0.004362026
 [5]  0.836891967  1.794961298  0.115284187  0.155175219
 [9]  0.464028612 -0.842569561
```

We could achieve a good visualization effect using R to draw a scatter diagram of the iris data set.

```
> data(iris) #Load data set.
> head(iris) #View the first 6 lines of the data set.
  Sepal.Length Sepal.Width Petal.Length Petal.Width Species
1          5.1         3.5          1.4         0.2  setosa
2          4.9         3.0          1.4         0.2  setosa
```

```
3         4.7         3.2         1.3         0.2   setosa
4         4.6         3.1         1.5         0.2   setosa
5         5.0         3.6         1.4         0.2   setosa
6         5.4         3.9         1.7         0.4   setosa

> plot(iris) # Draw the graphic.
```

The output result can be seen in Figure 1.1.

The free philosophy of R as well as its simple and special syntax are major reasons for its appeal.

1.1.2.7 Thinking Patterns of R

R has helped me clear away an old mindset. Using R to solve a problem, we should learn to think about questions from the perspective of statistics rather than from computing. In the thinking patterns of statistics, we usually consider why before we figure out what to do. But in the thinking patterns of computing, we just directly think about what to do and then consider the reason based on the result.

R is a programming language that directly deals with data. In our daily lives, data are produced in whatever we do, such as browsing data while surfing the Internet and consumption data in shopping. Even if we do nothing, our lives will be influenced by the air quality, and there will be data of air pollution indexes such as PM2.5. With the help of R, we can analyze these data directly.

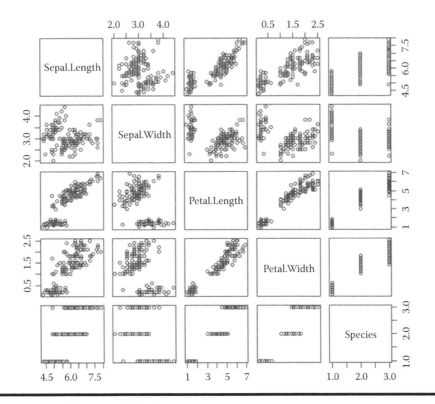

Figure 1.1 Scatter diagram of iris data set.

An advantage of the thinking patterns of R lies in the fact that we can simply analyze the data we need to cope with. We do not need to go through a role transition from programmer to product manager and consider what functions are involved, not to mention the program design.

Becoming free of the thinking patterns of programmers will help us learn more and find more suitable positions of ourselves.

1.1.2.8 Problems to Be Solved by R

Today people use data as a means of production. Thus R has become a productivity tool that can be used to deal with this new means of production and create value. Therefore R is mainly used to solve the problems of data.

The volume of data created in the era of the Internet exceeds that created in the entire long history of the pre-Internet age. How to realize the value of such a great amount of data has become the most popular topic as Hadoop helps people solve the problem of big data storage. The statistical analysis ability of R has thus made it the best tool for data analysis. Therefore, the problems to be solved by R are exactly the problems of this big data era.

1.1.2.9 Shortcomings of R

Although R has many merits, as discussed earlier, it does have some shortcomings.

- R, as a software created by statisticians, is not as robust as other software created by software engineers.
- Some problems exist in R's performance.
- R's essence of freedom somewhat denormalizes the naming of the syntax of Rand makes it difficult to familiarize.
- Some basic knowledge of mathematics, probability, and statistics is required to learn R.

However the difficulties of applying R caused by these shortcomings of R can be overcame. With more and more people with an engineering background joining the circle, R will become more powerful and help users to create more value in the future.

1.1.3 Application Prospects of R

R is able to accomplish all the assignments of Statistics Analysis System (SAS), which is one of the most famous commercial analysis software in the world. SAS is used as a large-scale integrated information system for decision support, so the statistics analysis function is an important component and core function of SAS. In the fields of data processing and statistics analysis, SAS is regarded as an international standard of a software system, or the giant in statistics software.

Now R is in a complete competitive relation with SAS. Being free and open, R will have a broader application prospect. The following are some major fields where R is applied.

- Statistics analysis: including statistical distribution, hypothesis testing, and statistical modeling
- Financial analysis: quantitative strategies, investment portfolio, risk control, time series, and volatility
- Data mining: data mining algorithms, data modeling, machine learning
- Internet: recommender system, consumption prediction, social network

- Bioinformatics: DNA analysis, species analysis
- Biopharmaceuticals: survival analysis, pharmaceutical process management
- Global geographic science: weather, climate, remote sensing data
- Data visualization: static diagram, interactive dynamic diagram, social diagram, map, heat map, and integration with all kinds of Java classes

I have written many articles concerning the application of R in my blog, including all the aforementioned fields except for biology. The broad application prospect of R has made it most capable of creating value in the new era.

1.1.4 Missions Assigned to R by the New Era

R is a programming language understood and known by the industry in this era of big data. Thus R has been given missions by the age, including exploring the value of data, finding patterns of data and creating fortunes by using data.

R is also an optimal productivity tool to help people make full use of their intelligence and creativity. We should not only learn to use R, but also try to raise the social productivity in all fields by applying R.

With all the elaboration of R, I believe R is the most worthwhile programming language to learn. Whether you are still at school or in a career, learning to use R will help you find the most suitable position and gain a bright future. In conclusion, among the five programming languages listed at the beginning of this chapter, R is the most special one and is tasked with different objectives. The generic functions of R determine that R will be a star programming language in 2014, or very possibly in the future.

1.2 Installation of Different Versions of R

Question

How do we install different versions of R in Linux Ubuntu?

The Installation of
Different Versions of R
http://blog.fens.me/r-install-ubuntu/

The R packages have progressed to version 3.2.2, but some of the third-party packages of R have still not been upgraded beyond version 2.15, such as RHadoop, RHive, and so forth. Thus certain versions of R are needed if we are to use these R packages.

This is a quite simple operation for Windows®: All we have to do is install different exe files. But those who are not familiar with Linux may have difficulty with the installation process. This section therefore describes the installation of certain versions of R packages in Windows and Linux Ubuntu®.

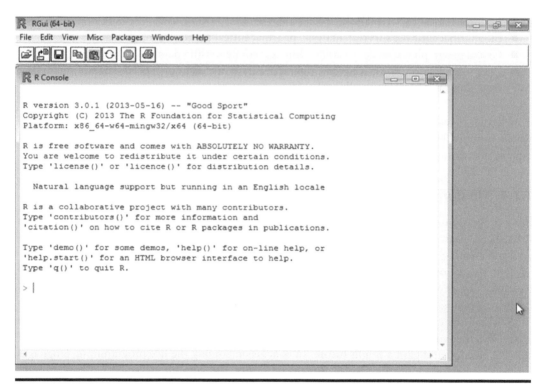

Figure 1.2   Installation of R in Windows®.

1.2.1 Installation of R in Windows

It is possible to download R packages of Linux, MacOS, and Windows from the official R website (http://cran.r-project.org/).

The installation of R in Windows is very easy. We just need to download the executable file and double-click it to install. The running interface of R after installation can be seen in Figure 1.2.

1.2.2 Installation of R in Linux Ubuntu

The Linux system used in this book is Ubuntu 12.04.2 LTS 64bit. The installation of R packages can be done by the tool apt-get in Ubuntu.

Installation of R in Linux Ubuntu

```
# Check whether R has been installed.
~ R
The program 'R' is currently not installed.  You can install it by typing:
sudo apt-get install r-base-core

# Install R packages under prompt
~ sudo apt-get install r-base-core
```

```
# Check the version of R.
~ R —version
R version 2.14.1 (2011-12-22)
Copyright (C) 2011 The R Foundation for Statistical Computing
ISBN 3-900051-07-0
Platform: x86_64-pc-linux-gnu (64-bit)
```

We can find that the default version of R is 2.14.1, which is different from the version used in this book. So next we wish to install the latest version of R packages.

1.2.3 Installation of Latest Version of R

First, delete the original R packages in Linux Ubuntu.

```
~ sudo apt-get autoremove r-base-core
```

Then, we find a mirror of software sources of Ubuntu (http://mirror.bjtu.edu.cn/cran/bin /linux/ubuntu/). The corresponding name of Linux Ubuntu 12.04 is precise. We can find different versions of R in r-base-care related files in precise/directory.

Add this software sources to sources.list of apt.

```
# Add a new line at the end of sources.list.
~ sudo sh -c "echo deb http://mirror.bjtu.edu.cn/cran/bin/linux/ubuntu precise
/>>/etc/apt/sources.list"

# Update the sources.
~ sudo apt-get update

# Re-install R packages.
~ sudo apt-get install r-base-core

# Check the version of R.
~ R —version
R version 3.0.3 (2014-03-06) — "Warm Puppy"
Copyright (C) 2014 The R Foundation for Statistical Computing
Platform: x86_64-pc-linux-gnu (64-bit)
```

Thus we have finished installing the latest version 3.0.3 of R.

1.2.4 Installation of Certain Versions of R

Because most of the cases in this book are completed in version 3.0.1 of R, while cases involving RHadoop are based on version 2.15.3, we need to install some specific versions of R.

1.2.4.1 Installation of Version 2.15.3 of R

```
# Delete the original R packages of the system.
~ sudo apt-get autoremove r-base-core

# Install the version 2.15.3 of R.
~ sudo apt-get install r-base-core=2.15.3-1precise0precise1

# Check the version of R.
~ R —version
R version 2.15.3 (2013-03-01) — "Security Blanket"
Copyright (C) 2013 The R Foundation for Statistical Computing
ISBN 3-900051-07-0
Platform: x86_64-pc-linux-gnu (64-bit)
```

1.2.4.2 Installation of Version 3.0.1 of R

```
# Delete the original R packages of the system.
~ sudo apt-get autoremove r-base-core

# Install the version 3.0.1 of R.
~ sudo apt-get install r-base-core=3.0.1-6precise0

# Check the version of R.
~ R —version
R version 3.0.1 (2013-05-16) — "Good Sport"
Copyright (C) 2013 The R Foundation for Statistical Computing
Platform: x86_64-pc-linux-gnu (64-bit)
```

Here we have installed the different versions of R conveniently to meet the different application demands.

1.3 fortunes: Records the Wisdom of R

Question

How do we get to know R in depth? What are the origin, growth, and experience of R?

The popularity of big data throughout the world has led to the gradual adoption of R. But because the R community has existed for many years, many of us might have little idea of the wisdom of R in its long history. Fortunately, someone is secretly recording the wisdom of R.

1.3.1 Introduction to fortunes

The fortunes library is a set of quotations of R. There were altogether 360 R-help messages up to December 14, 2013, which embody the essence of the wisdom of R. These messages make it easier for later R users to understand R and get to know the spirit of R.

1.3.2 Installation of fortunes

System environment used in this section:

- Linux: Ubuntu 12.04.2 LTS 64bit
- R:3.0.1 x86_64-pc-linux-gnu

Note: fortunes support both Windows 7 and Linux

Installation of fortunes:

```
# Launch R.
~ R

# Install fortunes packages.
> install.packages("fortunes")

# Load fortunes packages.
> library(fortunes)

# Check the help.
> ?fortunes
```

1.3.3 Use of fortunes

The use of fortunes is very simple. There is only one function in this package: fortune().

```
# Check a random quotation.
> fortune()
Barry Rowlingson: Your grid above has 8*6 = 42 points.
(That was a subtle Hitchhikers Guide To The Galaxy reference there, honest,
and not a stupid dumb multiplication mistake on my part after working four
18-hour days on the trot...)
Peter Dalgaard: [...] Don't panic, just throw yourself at the ground and miss.
    — Barry Rowlingson and Peter Dalgaard
      R-help (March 2004)

# Check a specific quotation.
> fortune(108)
Actually, I see it as part of my job to inflict R on people who are perfectly
happy to have never heard of it. Happiness doesn't equal proficient and
efficient. In some cases the proficiency of a person serves a greater good
than their momentary happiness.
    — Patrick Burns
      R-help (April 2005)
```

The complete quotations can be downloaded from http://cran.r-project.org/web/packages /fortunes/vignettes/fortunes.pdf.

We need to slow down and understand a language in depth if we wish to master it.

1.4 Using formatR to Format Codes Automatically

Question

How do we write codes that conform to specifications and can be easily understood by others?

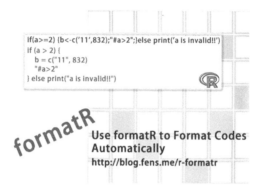

```
if(a>=2) {b<-c('11',832);"#a>2";}else print('a is invalid!!')
if (a > 2) {
    b = c("11", 832)
    "#a>2"
} else print("a is invalid!!")
```

Use formatR to Format Codes Automatically
http://blog.fens.me/r-formatr

Most beginners may focus only on how to realize certain functions when they write codes but ignore the importance of coding standards. Such codes will be unpopular not only to those who need to read them, but also to the writers themselves if they review their codes after a few months.

The most painful thing for a programmer is not working overtime every day to write codes, but working overtime every day to try to understand the programs written by others.

At first, most programmers did not consider how to make it more convenient for others to understand their codes. Finally someone started to make coding standards because he could no longer tolerate those ugly codes. Then someone invented tools that could implement the automatic formatting of codes. formatR is such a tool for automatic formatting of R.

1.4.1 Introduction to formatR

formatR is a useful R package that provides the function to format R codes. We can set the codes format of blank, retract, and line feed automatically in formatR, which may make the codes more user-friendly.

This is the introduction of API in formatR:

- tidy.source: to format codes
- tidy.eval: to output formatted R codes and the results
- usage: definition of format function and output with specific width
- tidy.guit: a GUI tool, which supports the editing and formatting of R codes
- tidy.dir: to format all the R scripts in a certain directory

1.4.2 Installation of formatR

System environment used in this section:

■ Windows 7 64bit
■ R: 3.0.1 x86_64-w64-mingw32/x64 b4bit

Note: formatR supports both Windows 7 and Linux.

Installation of formatR

```
# Launch R.
~ R

# Install formatR package.
> install.packages("formatR")
trying URL
'http://mirror.bjtu.edu.cn/cran/bin/windows/contrib/3.0/formatR_0.10.zip'
Content type 'application/zip' length 49263 bytes (48 Kb)
opened URL
downloaded 48 Kb

package 'formatR' successfully unpacked and MD5 sums checked

# Load formatR.
library(formatR)
```

1.4.3 Use of formatR

1.4.3.1 tidy.source: To Format Codes by Inputting a Character String

```
> tidy.source(text = c("{if(TRUE)1 else 2; if(FALSE){1+1", "## comments", "}
else 2}"))
{
    if (TRUE)
        1 else 2
    if (FALSE) {
        1 + 1
        ## comments
    } else 2
}
```

1.4.3.2 tidy.source: To Format Codes by Inputting Files

messy.R is a R file that does not conform to the standards well.

```
> messy = system.file("format", "messy.R", package = "formatR")
> messy
[1] "C:/Program Files/R/R-3.0.1/library/formatR/format/messy.R"
```

The original code output of messy.R:

```
> src = readLines(messy)
> cat(src,sep="\n")
    # a single line of comments is preserved
1+1

if(TRUE){
x=1  # inline comments
}else{
x=2;print('Oh no... ask the right bracket to go away!')}
1*3 # one space before this comment will become two!
2+2+2    # 'short comments'

lm(y~x1+x2, data=data.frame(y=rnorm(100),x1=rnorm(100),x2=rnorm(100))) ###
only 'single quotes' are allowed in comments
1+1+1+1+1+1+1+1+1+1+1+1+1+1+1+1+1+1+1+1 ## comments after a long line
'a character string with \t in it'

## here is a long long long long long long long long long long long long long
long long long long long long long comment
```

The code output after formatting:

```
> tidy.source(messy)
# a single line of comments is preserved
1 + 1

if (TRUE) {
    x = 1  # inline comments
}  else {
    x = 2
    print("Oh no... ask the right bracket to go away!")
}
1 * 3  # one space before this comment will become two!
2 + 2 + 2  # 'short comments'

lm(y ~ x1 + x2, data = data.frame(y = rnorm(100), x1 = rnorm(100), x2 =
rnorm(100))) ###  only 'single quotes' are allowed in comments
1 + 1 + 1 + 1 + 1 + 1 + 1 + 1 + 1 + 1 + 1 + 1 + 1 + 1 + 1 + 1 + 1 + 1 + 1
+ 1 + 1 + 1  ## comments after a long line
"a character string with \t in it"

## here is a long long long long long long long long long long long long long
long long long
## long long long long comment
```

The output after formatting has been processed with blank, retract, line feed, and comment, which increases the readability of the codes.

1.4.3.3 Formatting and Outputting R Script Files

Create a new R script file: demo.r:

```
~ vi demo.r

a<-1+1;a;matrix(rnorm(10),5);
if(a>2)  {b=c('11',832);"#a>2";}  else print('a is invalid!!')
```

Format demo.r:

```
> x = "demo.r"
> tidy.source(x)
a <- 1 + 1
a
matrix(rnorm(10), 5)
if (a > 2) {
    b = c("11", 832)
   "#a>2"
}  else print("a is invalid!!")
```

Output the result of formatting to file demo2.r, as in Figure 1.3.

```
> f="demo2.r"
> tidy.source(x, keep.blank.line = TRUE, file = f)
> file.show(f)
```

1.4.3.4 tidy.eval: Output Formatted R Codes and the Run Results

Run an R script in character string form:

```
> tidy.eval(text = c("a<-1+1;a", "matrix(rnorm(10),5)"))
a <- 1 + 1
a
## [1] 2

matrix(rnorm(10), 5)
##      [,1]       [,2]
## [1,]  0.65050729  0.1725221
## [2,]  0.05174598  0.3434398
## [3,] -0.91056310  0.1138733
## [4,]  0.18131010 -0.7286614
## [5,]  0.40811952  1.8288346
```

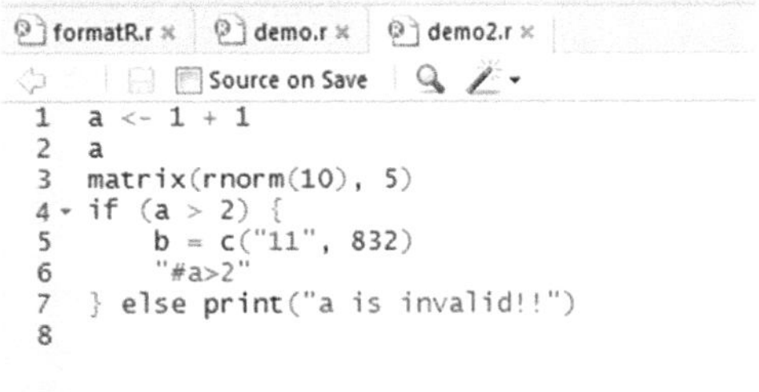

```
 formatR.r ×    demo.r ×    demo2.r ×
          Source on Save
1   a <- 1 + 1
2   a
3   matrix(rnorm(10), 5)
4 ▾ if (a > 2) {
5        b = c("11", 832)
6       "#a>2"
7   } else print("a is invalid!!")
8
```

Figure 1.3   Outputting the result of formatting to file.

1.4.3.5 usage: Definition of Format Function and Output with Specific Width

We can print out only the function definition and skip the details of function by using usage.

Take the var function as an example: the default would be printing out a function detail if we input var:

```
> var
function (x, y = NULL, na.rm = FALSE, use)
{
if (missing(use))
use <- if (na.rm)
"na.or.complete"
else "everything"
na.method <- pmatch(use, c("all.obs", "complete.obs", "pairwise.complete.obs",
"everything", "na.or.complete"))
if (is.na(na.method))
stop("invalid 'use' argument")
if (is.data.frame(x))
x <- as.matrix(x)
else stopifnot(is.atomic(x))
if (is.data.frame(y))
y <- as.matrix(y)
else stopifnot(is.atomic(y))
.Call(C_cov, x, y, na.method, FALSE)
}
<bytecode: 0x0000000008fad030>
<environment: namespace:stats>

# Print out only function definition by usage.
> usage(var)
var(x, y = NULL, na.rm = FALSE, use)
```

Sometimes the definition of a function can be very long, like in the function lm. We can control the display width of a function by the width parameter in usage.

```
> usage(lm)
lm(formula, data, subset, weights, na.action, method = "qr", model =
TRUE, x = FALSE, y = FALSE, qr = TRUE, singular.ok = TRUE, contrasts
= NULL, offset,...)

# Width parameter in usage, which can control the display width of a
function,
> usage(lm,width=30)
lm(formula, data, subset, weights,
    na.action, method = "qr", model = TRUE,
    x = FALSE, y = FALSE, qr = TRUE,
    singular.ok = TRUE, contrasts = NULL,
    offset,...)
```

1.4.3.6 tidy.gui: A GUI Tool Used to Edit and Format R Codes

tidy.gui() is a GUI tool that can be used to edit and format R codes in the interface. First install the gWidgetsRGtk2 library:

```
> install.packages("gWidgetsRGtk2")
also installing the dependencies 'RGtk2', 'gWidgets'

trying URL
'http://mirror.bjtu.edu.cn/cran/bin/windows/contrib/3.0/RGtk2_2.20.25.
zip'
Content type 'application/zip' length 13646817 bytes (13.0 Mb)
opened URL
downloaded 13.0 Mb

trying URL
'http://mirror.bjtu.edu.cn/cran/bin/windows/contrib/3.0/
gWidgets_0.0-52.zip'
Content type 'application/zip' length 1212449 bytes (1.2 Mb)
opened URL
downloaded 1.2 Mb

trying URL

'http://mirror.bjtu.edu.cn/cran/bin/windows/contrib/3.0/gWidgetsRGtk2_
0.0-82.zip'
Content type 'application/zip' length 787592 bytes (769 Kb)
opened URL
downloaded 769 Kb

package 'RGtk2' successfully unpacked and MD5 sums checked
package 'gWidgets' successfully unpacked and MD5 sums checked
package 'gWidgetsRGtk2' successfully unpacked and MD5 sums checked
```

Open GUI console:

```
> library("gWidgetsRGtk2")
> g = tidy.gui()
```

First we input a piece of code that is not so reader-friendly, as in Figure 1.4.

Click transform, and we can see that the R codes are formatted in GUI editor. The result can be seen in Figure 1.5.

1.4.3.7 tidy.fir: Format All the R Scripts in Directory dir

tidy.dir() can be used to format files in batch in certain directory. Now let's create a new directory: dir and create two R script files: dir.r, dir2.r in the directory dir.

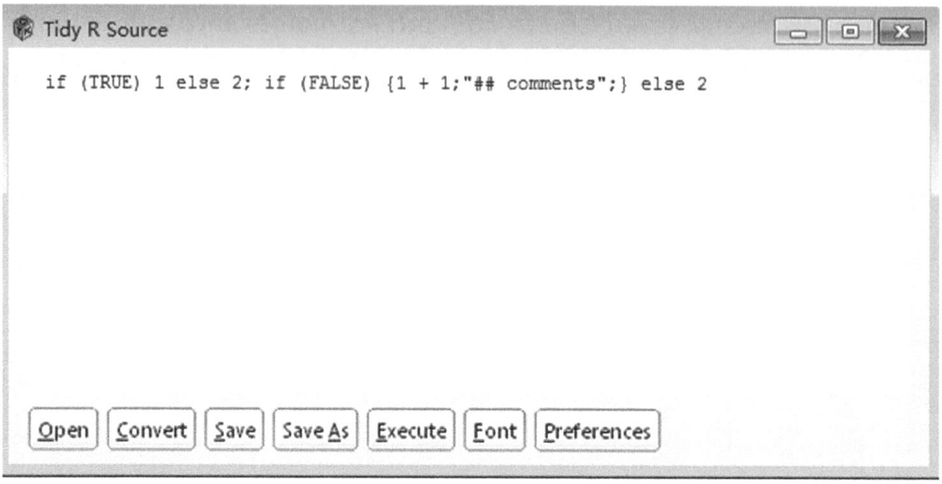

```
if (TRUE) 1 else 2; if (FALSE) {1 + 1;"## comments";} else 2
```

Figure 1.4   tidy in GUI.

```
if (TRUE) 1 else 2
if (FALSE) {
    1 + 1
    "## comments"
} else 2
```

Figure 1.5   Codes after formatting.

```
# Create directory dir.
~ mkdir dir
~ cd dir

# Use vi to create file dir.r
~ vi dir.r
a<-1+1;a;matrix(rnorm(10),5);

~ vi dir2.r
if(a>2) {b=c('11',832);"#a>2";} else print('a is invalid!!')
```

Run tidy.dir:

```
> tidy.dir(path="dir")
tidying dir/dir.r
tidying dir/dir2.r
```

View dir.r and dir2.r:

```
~ vi dir.r
a <- 1 + 1
a
matrix(rnorm(10), 5)

~ vi dir2.r
if (a > 2) {
    b = c("11", 832)
    "#a>2"
}  else print("a is invalid!!")
```

The codes have already been formatted!

1.4.4 Source Code Analysis of formatR

After all of the preceding operations, we easily find that the core function of formatR is tidy.source. The source code can be downloaded from Github: https://github.com/yihui/formatR/blob /master/R/tidy.R.

I have added comments to the code:

```
tidy.source = function(
  source = 'clipboard', keep.comment = getOption('keep.comment', TRUE),
  keep.blank.line = getOption('keep.blank.line', TRUE),
  replace.assign = getOption('replace.assign', FALSE),
  left.brace.newline = getOption('left.brace.newline', FALSE),
  reindent.spaces = getOption('reindent.spaces', 4),
  output = TRUE, text = NULL,
  width.cutoff = getOption('width'),...
) {

  # Determine the source of input is clipboard.
  if (is.null(text)) {
    if (source == 'clipboard' && Sys.info()['sysname'] == 'Darwin') {
      source = pipe('pbpaste')
    }
```

```
  } else {  # Determine the source of input is character string.
    source = textConnection(text); on.exit(close(source))
  }

  # Read source data by lines.
  text = readLines(source, warn = FALSE)

  # Length processing.
  if (length(text) == 0L || all(grepl('^\\s*$', text))) {
    if (output) cat('\n',...)
    return(list(text.tidy = text, text.mask = text))
  }

  # Blank processing.
  if (keep.blank.line && R3)  {
    one = paste(text, collapse = '\n') # record how many line breaks
before/after
    n1 = attr(regexpr('^\n*', one), 'match.length')
    n2 = attr(regexpr('\n*$', one), 'match.length')
  }

  # Comment processing.
  if (keep.comment) text = mask_comments(text, width.cutoff, keep.
blank.line)

  # Transform the R codes into expression and back to character
string,to implement the interception of each statement.
  text.mask = tidy_block(text, width.cutoff, replace.assign &&
length(grep('=', text)))

  # Format the comments.
  text.tidy = if (keep.comment) unmask.source(text.mask) else text.
mask

  # Re-locate the retract.
  text.tidy = reindent_lines(text.tidy, reindent.spaces)

  # Linefeed of brackets.
  if (left.brace.newline) text.tidy = move_leftbrace(text.tidy)

  # Add null string to the beginning and the end.
  if (keep.blank.line && R3) text.tidy = c(rep('', n1), text.tidy,
rep('', n2))

  # Print out the formatted results in console.
  if (output) cat(paste(text.tidy, collapse = '\n'), '\n',...)

  # Return without printing out the results.
  invisible(list(text.tidy = text.tidy, text.mask = text.mask))
```

1.4.5 *Bugs in the Source Code*

I find a small problem while reading the source code: Assigning toward the right has not been processed in version 3.0.1 of R. The bug has been reported to the author at https://github.com /yihui/formatR/issues/31.

Code of bug testing:

```
> c('11',832)->x2
> x2
[1] "11"  "832"

# Format the code.
> tidy.source(text="c('11',832)->x2")
c("11", 832) <- x2

> tidy.eval(text="c('11',832)->x2")
c("11", 832) <- x2
Error in eval(expr, envir, enclos): object 'x2' not found
```

The bug has been fixed. The author's reply: This bug has been fixed in R 3.0.2.

```
# Format the code.
> formatR::tidy.source(text="c('11',832)->x2")
x2 <- c("11", 832)
> sessionInfo()
R version 3.0.2 (2013-09-25)
Platform: x86_64-pc-linux-gnu (64-bit)

locale:
 [1] LC_CTYPE=en_US.UTF-8       LC_NUMERIC=C
 [3] LC_TIME=en_US.UTF-8        LC_COLLATE=en_US.UTF-8
 [5] LC_MONETARY=en_US.UTF-8    LC_MESSAGES=en_US.UTF-8
 [7] LC_PAPER=en_US.UTF-8       LC_NAME=C
 [9] LC_ADDRESS=C               LC_TELEPHONE=C
[11] LC_MEASUREMENT=en_US.UTF-8 LC_IDENTIFICATION=C

attached base packages:

[1] stats     graphics  grDevices utils     datasets  methods   base

loaded via a namespace (and not attached):
[1] formatR_0.10.3
```

The functions provided in formatR are very useful and practical, especially when we read some codes that do not conform to the standards very well. Here I suggest that integrated development environment (IDE) insert formatR into the tools of editors as a standardized formatting tool. We hope that reading others' codes may become a happy experience in the future.

1.5 Multiuser Online Collaboration of R Development: RStudio Server

Question

Which tool is the best for R development?

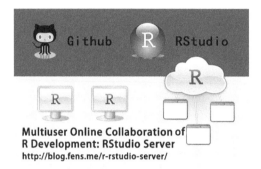

Multiuser Online Collaboration of R Development: RStudio Server
http://blog.fens.me/r-rstudio-server/

RStudio is an effective tool for R development, and it's also the best IDE integrated environment for R. RStudio Server is one of the best in RStudio. It not only provides Web functions, which can be installed on remote server and visited by the Web, but also supports multiuser collaboration development. Let's try to use such an effective tool!

1.5.1 RStudio and RStudio Server

RStudio is a powerful, free and open application software of integrated development environment of R. It can be installed on many systems including Windows, Linux, and Mac OS. RStudio Server, a cloud development environment of RStudio based on Web visiting, needs to be installed on a Linux server to support remote multiuser visits.

1.5.2 Installation of RStudio Server

The system environment used in this section:

- Linux: Ubuntu Server 12.04.2 LTS 64bit
- R: 3.0.1 x86_64-pc-linux-gnu
- IP:192.168.1.13

Note: RStudio Server supports only the Linux system. The latest version of RStudio Server can be downloaded from http://www.rstudio.com/ide/download/server.html/.

Download and install RStudio Server of 64bit in Linux Ubuntu:

```
~ sudo apt-get install gdebi-core
~ sudo apt-get install libapparmor1  # Required only for Ubuntu, not
Debian
~ wget http://download2.rstudio.org/rstudio-server-0.97.551-amd64.
deb
~ sudo gdebi rstudio-server-0.97.551-amd64.deb
```

RStudio Server will be launched automatically after installation.

```
# Check the running process of RStudio Server.
~ ps -aux|grep rstudio-server
998       2914  0.0  0.1 192884  2568 ?          Ssl  10:40   0:00 /
usr/lib/rstudio-server/bin/rserver
```

It can be seen that RStudio Server has been launched, and port 8787 has been opened.

1.5.3 Use of RStudio Server

We can visit RStudio Server at http://192.168,1,13:8787, as in Figure 1.6.

A user account for the Linux system is needed if we log in to RStudio Server. We could operate directly on the Linux system if we need to increase or decrease the number of users. In this case, the username is conan, and the password is conan111. The screen after log-in can be seen in Figure 1.7.

1.5.3.1 System Configuration of RStudio Server

There are two main configuration files of RStudio Server. The default file does not exist.

```
/etc/rstudio/rserver.conf
/etc/rstudio/rsession.conf
```

Figure 1.6   Log-in screen of RStudio Server.

Figure 1.7   Web screen of RStudio Server.

Set up the port and ip control:

```
~ vi/etc/rstudio/rserver.conf

# Listener port.
www-port=8080

# IP address allowed for visiting. The default is 0.0.0.0.
www-address=127.0.0.1
```

Restart RStudio Server and the configuration have taken effect.

```
~ sudo rstudio-server restart
```

Management of session configuration.

```
~ vi/etc/rstudio/rsession.conf

# Timeout minutes of session.
session-timeout-minutes=30

# Setting up CRAN repository.
r-cran-repos=http://ftp.ctex.org/mirrors/CRAN/
```

1.5.3.2 System Management of RStudio Server

Here are the commands to start, stop, and restart RStudio Server.

```
# Start
~ sudo rstudio-server start

# Stop
~ sudo rstudio-server stop

# Restart
~ sudo rstudio-server restart
```

Check that the R process is running.

```
~ sudo rstudio-server active-sessions

PID     TIME COMMAND
6817 00:00:03/usr/lib/rstudio-server/bin/rsession -u zd
```

Point PID, and stop the R process from running.

```
~ sudo  rstudio-server suspend-session 6817

# Check the process again.
~ sudo rstudio-server active-sessions
    PID     TIME COMMAND
```

Suspend all the R processes from running.

```
~ sudo rstudio-server suspend-all
```

Force suspension of the running of the R process. This is an operation with the highest priority, which will be executed immediately.

```
~ sudo rstudio-server force-suspend-session <pid>
~ sudo rstudio-server force-suspend-all
```

A temporary offline of RStudio Server will reject Web visiting and give users a friendly error.

```
~ sudo rstudio-server offline
rstudio-server start/running, process 6880
```

Online of RStudio Server.

```
~ sudo rstudio-server online
rstudio-server start/running, process 6908
```

Other operations of RStudio Server are just the same as in the standalone version of RStudio.

1.5.4 Multiuser Collaboration of RStudio Server

1.5.4.1 Add New Users and New User Groups

```
# Create user group of Hadoop.
~ sudo groupadd hadoop

# Create Hadoop user and add it to user group.
~ sudo useradd hadoop -g hadoop

# Set the password for Hadoop user.
~ sudo passwd hadoop

# Add Hadoop user to sudo.
~ sudo adduser hadoop sudo

# Create home directory for Hadoop user.
~ sudo mkdir/home/hadoop

# Set user permission for/home/Hadoop directory and its subdirectory.
~ sudo chown -R hadoop:hadoop/home/hadoop
```

Log in as a Hadoop user to check whether the setup is successful.

```
# Remote log in through ssh.
~ ssh hadoop@localhost
~ bash

# View the visiting directory after logging in.
~ pwd
/home/hadoop
```

Open a new browser window and log in through the Hadoop account, as in Figure 1.8.

Figure 1.8   Log in through Hadoop account.

1.5.4.2 Share of Git Codes

First, install Git.

```
sudo apt-get install git

# Generate rsa key pair.
ssh-keygen -t rsa

# View the public key.
cat/home/conan/.ssh/id_rsa.pub
ssh-rsa

AAAAB3NzaC1yc2EAAAADAQABAAABAQDMmnFyZe2RHpXaGmENdH9kSyDyVzRas4GtRwMN
x+qQ4QsB8xVTrIbFayG2ilt+P8UUkVYO0qtUJIaLRjGy/SvQzzL7JKX12+VyYoKTfKvZ
ZnANJ414d6oZpbDwsC0Z7JARcWsFyTW1KxOMyesmzNNdB+F3bYN9sYNiTkOeVNVYmEQ8
aXywn4kc1jBhVpT8PbuHl5eadSLt5zpN6bcX7tlquuTlRpLi1e4K+8j
Qo67H54FuDyrPLUYtVaiTNT/xWN6IU+DQ9CbfykJ0hrfDU1d1LiLQ4K2Fdg+vcKtB7Wxe
z2wKjsxb4Cb8TLSbXdIKEwSOFooINw25g/Aamv/nVvW1 conan@conan-deskop
```

Then we should upload the local project to Github. Create a new project rstudio-demo on Github, with the address https://github.com/bsspirit/rstudio-demo. Upload the local directory to rstudio-demo project by the operations that follow.

```
# Create directory for rstudio-demo project.
~ mkdir /home/conan/R/github
~ cd /home/conan/R/github

# Initialize git.
~ git init

# Add current directory and its subdirectory to local Git library.
~ git add.

# Commit in local Git library.
~ git commit -m 'first comment'

# Bind current directory with Github projetct.
~ git remote add origin git@github.com:bsspirit/rstudio-demo.git

# Upload codes in local git library to Github.
~ git push -u origin masterv
```

Open RStudio and set it to directory/home/conan/R/Github, tools –> version control –> project setup, as in Figure 1.9.

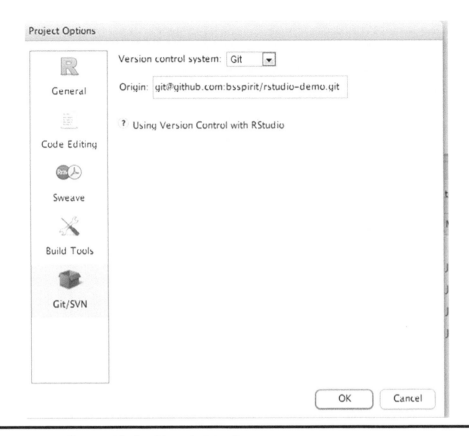

Figure 1.9   Configure Github address in RStudio.

Modify the codes of sayHello.r in RStudio:

```
sayHello<-function(name){
  print(paste("hello",name))
}

sayHello("Conan")
sayHello("World")
```

Commit: click tools –> version control –> commit, as in Figure 1.10.

Upload to Github: click tools –> version control –> push, as in Figure 1.11.

These powerful functions of RStudio have made programming quite easy to learn. Let's start doing this and try to be a real geek!

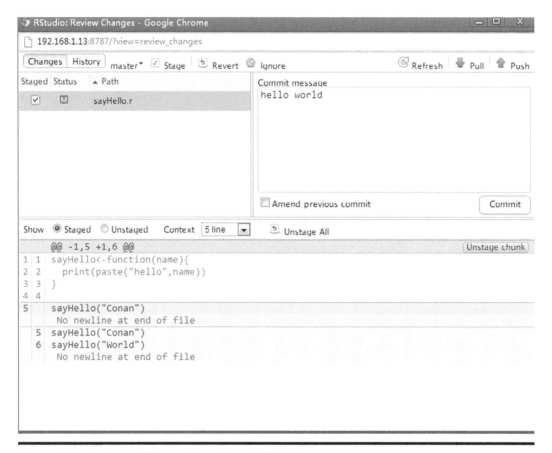

Figure 1.10   Perform Git operations through RStudio.

1.6 Foolproof Programming of R and JSON

Question

How do we convert an R data class into a JSON data class?

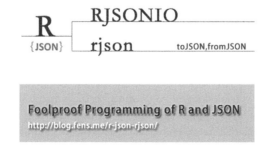

As a lightweight data format, JSON (JavaScript Object Notation) has been widely applied in all kinds of environments. JSON is a standard object embedded in JavaScript, and a storage class of

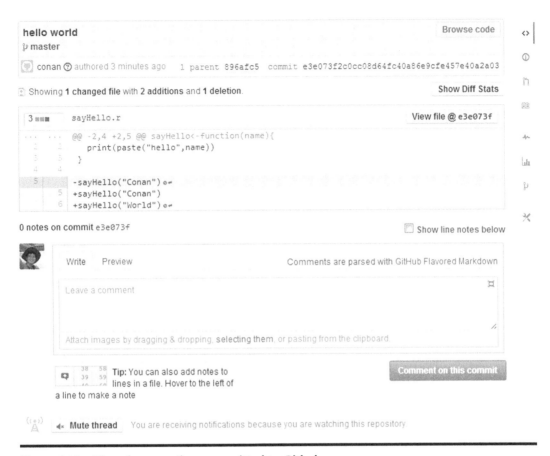

Figure 1.11 View the operations committed to Github.

table structure of MongoDB as well. JSON is semistructured, and it can express a rich meaning of documents. There are fewer JSON documents than XML documents, which makes JSON documents more suitable for network transmission. As it is at an early stage, JSON is rarely used in R programming. But as R becomes more popular and powerful, it has extended to many other fields, including JSON. How can one convert a JSON data class into an R data class in a foolproof way? The following is an introduction.

1.6.1 Introduction to rjson

rjson is an R package to make conversions between R and JSON that is very simple and supports two ways of converting: based on R itself and based on the C library. Rjson provides only three functions: fromJSON(), newJSONParser(), toJSON(). Here we introduce how to use rjson. The system environment used in this section is

■ Win7: x86_64-w64-mingw32/x64 (64-bit)
■ R: version 3.0.1

Note: rjson supports both Windows 7 and Linux.

1.6.1.1 Install and Load rjson

```
> install.packages("rjson")
> library(rjson)
```

Then we create a new JSON file for testing.
Create JOSN file: fin0.json:

```
~ vi fin0.json

{
    "table1": {
        "time": "130911",
        "data": {
            "code": [
                "TF1312",
                "TF1403",
                "TF1406"
            ],
            "rt_time": [
                130911,
                130911,
                130911
            ]
        }
    },
    "table2": {
        "time": "130911",
        "data": {
            "contract": [
                "TF1312",
                "TF1312",
                "TF1403"
            ],
            "jtid": [
                99,
                65,
                21
            ]
        }
    }
}
```

1.6.1.2 Call Function: fromJSON(): from JSON to R

Read JSON and parse it into an R object from fin0.json. We usually name the process of converting a byte or text formatting into a program an object deserialization process. The opposite process is termed a serialization process.

```
> json_data <- fromJSON(paste(readLines("fin0.json"), collapse=""))
> json_data
$table1
$table1$time
[1] "130911"

$table1$data
$table1$data$code
[1] "TF1312" "TF1403" "TF1406"

$table1$data$rt_time
[1] 130911 130911 130911

$table2
$table2$time
[1] "130911"

$table2$data
$table2$data$contract
[1] "TF1312" "TF1312" "TF1403"

$table2$data$jtid
[1] 99 65 21
```

Check the data class of R after converting:

```
> class(json_data)
[1] "list"

> class(json_data$table2)
[1] "list"

> class(json_data$table2$data)
[1] "list"

> class(json_data$table2$data$jtid)
[1] "numeric"

> class(json_data$table1$data$code)
[1] "character"
```

It can be seen that after converting, the original JSON object has been parsed into an R list except for the innermost part, which is now a basic class (numeric, character). If you take a leaf node of JSON data in R object structure, the index path of JSON will be json.table1. data.code[0].

```
> json_data$table1$data$code
[1] "TF1312" "TF1403" "TF1406"

> json_data$table1$data$code[1]
[1] "TF1312"
```

1.6.1.3 toJSON(): from R to JSON

The process of turning R object into JSON string is named serialization process. Again we use json_data above as example.

```
> json_str<-toJSON(json_data)

> print(json_str)
[1]
"{\"table1\":{\"time\":\"130911\",\"data\":{\"code\":[\"TF1312\",\"T
F1403\",\"TF1406\"],\"rt_time\":[130911,130911,130911]}},\"table2\":
{\"time\":\"130911\",\"data\":{\"contract\":[\"TF1312\",\"TF1312\",\
"TF1403\"],\"jtid\":[99,65,21]}}}"

> cat(json_str)
{"table1":{"time":"130911","data":{"code":["TF1312","TF1403","TF1406
"],"rt_time":[130911,130911,130911]}},"table2":{"time":"130911","dat
a":{"contract":["TF1312","TF1312","TF1403"],"jtid":[99,65,21]}}}
```

We can implement the conversion from an R object to JSON if we use the function toJSON(). If we output the result with the function print(), it would be an escaping output(\"). If we output the result with the function cat, it would be a standard JSON string. There are two ways to output JSON to fin0_out.json: writeLines() and skin().

```
# writeLines
> writeLines(json_str, "fin0_out1.json")

# sink
> sink("fin0_out2.json")
> cat(json_str)
> sink()
```

Although the code is different, the output will be the same. WriteLines will create a blank line at the end.

```
{"table1":{"time":"130911","data":{"code":["TF1312","TF1403","TF1406
"],"rt_time":[130911,130911,130911]}},"table2":{"time":"130911","dat
a":{"contract":["TF1312","TF1312","TF1403"],"jtid":[99,65,21]}}}
```

1.6.1.4 Converting between the C Library and R Library, and Performance Test

Run a performance test to fromJSON:

```
> system.time(y <- fromJSON(json_str,method="C"))
User System Elapse
    0    0    0
> system.time(y2 <- fromJSON(json_str,method = "R") )
User System Elapse
0.02 0.00 0.02
> system.time(y3 <- fromJSON(json_str))
User System  Elapse
    0    0    0
```

It can be seen that an operation based on the C library is faster than that based on the R library. The difference of 0.02 is not so obvious because the amount of data is small. When the JSON string gets much bigger, the difference will be bigger too. The default way adopted by fromJSON is C, so we do not need to add the parameter method = 'C' normally. Next we run a performance test of toJSON.

```
> system.time(y <- fromJSON(json_str,method="C") )
User System Elapse
    0    0    0
> system.time(y2 <- fromJSON(json_str,method = "R"))
User System Elapse
0.02 0.00 0.01
> system.time(y3 <- fromJSON(json_str))
User System  Elapse
    0    0    0
```

The explanation is the same as given previously.

1.6.2 Introduction to RJSONIO

RSJONIO provides two main operations: deserialize a JSON string into an R object, and serialize an R object into a JSON string. The two main functions of RJSONIO are fromJSON() and toJSON(). There are other helper functions: asJSVars(), basicJSONHandler(), isValidJSON() and readJSON-Stream(). It is relatively slow to serialize a large object in rjson, whereas RJSONIO has solved this problem. RJSONIO depends on the underlying C library, libjson.

1.6.2.1 Install and Load RJSONIO

```
> install.packages("RJSONIO")
> library(RJSONIO)
```

1.6.2.2 fromJSON(): from JSON to R

Test function fromJSON as in rjson.

```
> json_data <- fromJSON(paste(readLines("fin0.json"), collapse=""))
> json_data
$table1
$table1$time
[1] "130911"

$table1$data
$table1$data$code
[1] "TF1312" "TF1403" "TF1406"

$table1$data$rt_time
[1] 130911 130911 130911

$table2
$table2$time
[1] "130911"

$table2$data
$table2$data$contract
[1] "TF1312" "TF1312" "TF1403"

$table2$data$jtid
[1] 99 65 21
```

We find that the result is the same as in rjson: The class of R object is all lists except for the innermost part. Then take a leaf node:

```
> json_data$table1$data$code
[1] "TF1312" "TF1403" "TF1406"

> json_data$table1$data$code[1]
[1] "TF1312"
```

1.6.2.3 toJSON: from R to JSON

Run a performance test for toJSON:

```
> json_str<-toJSON(json_data)

> print(json_str)
[1] "{\n \"table1\": {\n \"time\": \"130911\",\n\"data\": {\n
\"code\": [\"TF1312\", \"TF1403\", \"TF1406\"],\n\"rt_time\":
```

```
[1.3091e+05, 1.3091e+05, 1.3091e+05] \n} \n},\n\"table2\": {\n
\"time\": \"130911\",\n\"data\": {\n \"contract\": [\"TF1312\",
\"TF1312\", \"TF1403\"],\n\"jtid\": [      99,      65,      21 ] \n}
\n} \n}"

> cat(json_str)
{
 "table1": {
 "time": "130911",
"data":  {
 "code": [ "TF1312", "TF1403", "TF1406" ],
"rt_time": [ 1.3091e+05, 1.3091e+05, 1.3091e+05 ]
}
},
"table2":  {
 "time": "130911",
"data":  {
 "contract": [ "TF1312", "TF1312", "TF1403" ],
"jtid": [      99,      65,      21]
}
}
}
```

The output of toJSON is formatted, which is different from that of rjson. Then output it to the file:

```
> writeLines(json_str, "fin0_io.json")
```

The result:

```
{
 "table1":  {
 "time": "130911",
 "data":  {
 "code": [ "TF1312", "TF1403", "TF1406" ],
 "rt_time": [ 1.3091e+05, 1.3091e+05, 1.3091e+05 ]
 }
 },
 "table2":  {
 "time": "130911",
 "data": {
 "contract": [ "TF1312", "TF1312", "TF1403"],
 "jtid": [      99,      65,      21 ]
 }
 }
}
```

1.6.2.4 *isValidJSON(): Check Whether JSON Is Valid*

Check whether the format of JSON is valid.

```
> isValidJSON(json_str)
Error in file(con, "r"): cannot open the connection

> isValidJSON(json_str,TRUE)
[1] TRUE

> isValidJSON(I('{"foo": "bar"}'))
[1] TRUE
> isValidJSON(I('{foo: "bar"}'))
[1] FALSE
```

1.6.2.5 *asJSVars(): Convert into a Variable Format of JavaScript*

```
> cat(asJSVars( a = 1:10, myMatrix = matrix(1:15, 3, 5)))
a = [1, 2, 3, 4, 5, 6, 7,f  8, 9, 10 ] ;

myMatrix = [ [ 1, 4, 7, 10, 13 ],
             [ 2, 5, 8, 11, 14 ],
             [ 3, 6, 9, 12, 15 ] ] ;
```

We get two JavaScript variables, array a and two-dimensional array myMatrix.

1.6.3 Implementation of Customized JSON

Data.frame is the most frequently used class in R. Next we will implement the conversion between data.frame and JSON. First, convert a data.frame object of R into a JSON format defined by us.
Define the output format of JSON:

```
[
    {
        "code": "TF1312",
        "rt_time": "152929",
        "rt_latest": 93.76,
        "rt_bid1": 93.76,
        "rt_ask1": 90.76,
        "rt_bsize1": 2,
        "rt_asize1": 100,
        "optionValue": -0.4,
        "diffValue": 0.6
    }
]
```

Define a data.frame:

```
> df<-data.frame(
+   code=c('TF1312','TF1310','TF1313'),
+   rt_time=c("152929","152929","152929"),
+   rt_latest=c(93.76,93.76,93.76),
+   rt_bid1=c(93.76,93.76,93.76),
+   rt_ask1=c(90.76,90.76,90.76),
+   rt_bsize1=c(2,3,1),
+   rt_asize1=c(100,1,11),
+   optionValue=c(-0.4,0.2,-0.1),
+   diffValue=c(0.6,0.6,0.5)
+)

> df
     code rt_time rt_latest rt_bid1 rt_ask1 rt_bsize1 rt_asize1
optionValue diffValue
1 TF1312  152929     93.76   93.76   90.76         2       100
-0.4        0.6
2 TF1310  152929     93.76   93.76   90.76         3         1
0.2         0.6
3 TF1313  152929     93.76   93.76   90.76         1        11
-0.1        0.5
```

Use toJSON directly. The JSON string was transformed to array by its columns, which is not what we want.

```
> cat(toJSON(df))
{
 "code": [ "TF1312", "TF1310", "TF1313" ],
"rt_time": [ "152929", "152929", "152929" ],
"rt_latest": [  93.76,   93.76,   93.76 ],
"rt_bid1": [  93.76,   93.76,   93.76 ],
"rt_ask1": [  90.76,   90.76,   90.76 ],
"rt_bsize1": [      2,       3,       1 ],
"rt_asize1": [    100,       1,      11 ],
"optionValue": [    -0.4,     0.2,    -0.1 ],
"diffValue": [    0.6,     0.6,     0.5 ]
}
```

We need to run data processing to data.frame:

```
# Use plyr to run data processing
> library(plyr).
> cat(toJSON(unname(alply(df, 1, identity))))
[
 {
 "code": "TF1312",
```

```
"rt_time": "152929",
"rt_latest":   93.76,
"rt_bid1":   93.76,
"rt_ask1":   90.76,
"rt_bsize1":       2,
"rt_asize1":     100,
"optionValue":    -0.4,
"diffValue":     0.6
},
{
 "code": "TF1310",
"rt_time": "152929",
"rt_latest":   93.76,
"rt_bid1":   93.76,
"rt_ask1":   90.76,
"rt_bsize1":       3,
"rt_asize1":       1,
"optionValue":     0.2,
"diffValue":     0.6
},
{
 "code": "TF1313",
"rt_time": "152929",
"rt_latest":   93.76,
"rt_bid1":   93.76,
"rt_ask1":   90.76,
"rt_bsize1":       1,
"rt_asize1":      11,
"optionValue":    -0.1,
"diffValue":     0.5
}
]
```

The output is displayed through data conversion by the function alply(). Such output by lines is exactly what we want.

1.6.4 Performance Comparison of JSON

We will run two performance tests: One is a serialization test (toJSON) on a large object between rjson and RJONIO and the other is a serialization test (toJSON) between the output by lines and output by columns in RJSONIO.

1.6.4.1 Serialization Test (toJSON) on a Large Object between rjson and RJSONIO

Create a test script of rjson and run it on the command line.

```
> library(rjson)
> df<-data.frame(
```

```
+    a=rep(letters,10000),
+    b=rnorm(260000),
+    c=as.factor(Sys.Date()-rep(1:260000))
+ )
>
> system.time(rjson::toJSON(df))
1.01 0.02 1.03
> system.time(rjson::toJSON(df))
1.01 0.03 1.04
> system.time(rjson::toJSON(df))
0.98 0.05 1.03
```

And create a test script of RJSONIO and run it on the command line.

```
> library(RJSONIO)
> df<-data.frame(
+    a=rep(letters,10000),
+    b=rnorm(260000),
+    c=as.factor(Sys.Date()-rep(1:260000))
+ )

> system.time(RJSONIO::toJSON(df))
2.23 0.02 2.24
> system.time(RJSONIO::toJSON(df))
2.30 0.00 2.29
> system.time(RJSONIO::toJSON(df))
2.25 0.01 2.26
```

The results of a comparison show that the performance of rjson is better than that of RJSONIO.

1.6.4.2 Serialization Test (toJSON) between Output by Lines and Output by Columns in RJSONIO

Create a test script of rjson and run it on the command line.

```
> library(rjson)
> library(plyr)

> df<-data.frame(
+    a=rep(letters,100),
+    b=rnorm(2600),
+    c=as.factor(Sys.Date()-rep(1:2600))
+ )

> system.time(rjson::toJSON(df))
0.01 0.00 0.02
> system.time(rjson::toJSON(df))
0.01 0.00 0.02
```

```
> system.time(rjson::toJSON(unname(alply(df, 1, identity))))
1.55 0.02 1.56
> system.time(rjson::toJSON(unname(alply(df, 1, identity))))
0.83 0.00 0.83
```

And create a test script of RJSONIO and run it on the command line.

```
> library(RJSONIO)
> library(plyr)

> df<-data.frame(
+   a=rep(letters,100),
+   b=rnorm(2600),
+ c=as.factor(Sys.Date()-rep(1:2600))
+ )
>
> system.time(RJSONIO::toJSON(df))
0.03 0.00 0.03
> system.time(RJSONIO::toJSON(df))
0.04 0.00 0.03

> system.time(RJSONIO::toJSON(unname(alply(df, 1, identity))))
2.82 0.02 2.84
> system.time(RJSONIO::toJSON(unname(alply(df, 1, identity))))
2.06 0.00 2.06
```

The result shows that output by columns is much more efficient than output by lines. Obviously output by lines will run through extra data processes.

The test result that rjson is more efficient than RJSONIO cannot support the introduction of RJSONIO. Of course, my one test cannot fully prove this conclusion. I hope that you can run some tests by yourselves if you have enough time and interest.

1.7 High-Quality Graphic Rendering Library of R Cairo

Question

How do we draw a high-definition graphic without jagged edges using R?

R not only offers strong computing power in statistical analysis and data mining, but also qualifies as a powerful tool in data visualization when compared to other expensive commercial software. However, R cannot be strong in visualization without the support of all kinds of open source packages. Cairo is such a class library used in vector graphic processing. You can create high-quality vector graphics (GIF, SVG, PDF, PostScript) and bitmaps (PNG, JEPG, TIFF) in Cairo and run high-quality rendering in background programs as well. This section introduces how to use Cairo in R.

1.7.1 Introduction to Cairo

Cairo is a free library used for graphic drawing and rendering. It supports complex 2D drawing and hardware acceleration. Although Cairo is written in C language, it provides an interface for many languages and allows other languages to call it directly, including C++, C#, Java, Python, Perl, Ruby, Scheme, and Smalltalk. The license agreement published by Cairo is GNU Lesser General Public License version 2.1 (LGPL), or Mozilla Public License 1.1 (MPL). The official release page of Cairo's interface for R is http://www.rforge.net/Cairo/.

1.7.2 Installation of Cairo

System environment used in this section:

- Linux: Ubuntu 12.04.2 LTS 64bit
- R: 3.0.1 x86_64-pc-linux-gnu

Note: Cairo supports both Windows 7 and Linux.

Installation of Cairo in Linux Ubuntu is as follows.

```
# Underlying dependant libraries of Cairo.
~ sudo apt-get install libcairo2-dev
~ sudo apt-get install libxt-dev

# Start R.
~ R

# Install Cairo.
> install.packages("Cairo")
```

1.7.3 Use of Cairo

It is very easy to use Cairo. It corresponds to the functions of the basic package grDevices.

- CairoPNG: grDevices:png()
- CairoJPEG: grDevices:jepg()
- CairoTIFF: grDevices:tiff()
- CairoSVG: grDevices:svg()
- Cairo PDF: grDevices:pdf()

The usual graphic output of me is png and svg. Here let's check the compatibility of Cairo:

```
# Load Cairo.
> library(Cairo)

# Check the capability of Cairo.
> Cairo.capabilities()
   png   jpeg   tiff    pdf    svg     ps    x11    win raster
  TRUE   TRUE  FALSE   TRUE   TRUE   TRUE   TRUE  FALSE   TRUE
```

Here is the result of this check:

■ Support: png, jpeg, pdf, svg, ps, xl1 (Linux desktop), raster
■ Not support: tiff, win (win desktop)

Note: x11 would be FALSE and win would be TRUE in Windows.

Then let's compare the effect of output by CairoPNG() and png().

1.7.3.1 Scatterplot

First we draw a scatterplot with 6000 points.

```
# Pick 6000 point coordinates randomly.
> x<-rnorm(6000)
> y<-rnorm(6000)

# png function.
> png(file="plot4.png",width=640,height=480)
> plot(x,y,col="#ff000018",pch=19,cex=2,main = "plot")
> dev.off()

# CairoPNG function
> CairoPNG(file="Cairo4.png",width=640,height=480)
> plot(x,y,col="#ff000018",pch=19,cex=2,main = "Cairo")
> dev.off()
```

Two png files, plot4.pgn and Cairo4.png, will be generated in the current directory, as in Figures 1.12 and 1.13.

The code of graphic output of SVG is

```
> svg(file="plot-svg4.svg",width=6,height=6)
> plot(x,y,col="#ff000018",pch=19,cex=2,main = "plot-svg")
> dev.off()

> CairoSVG(file="Cairo-svg4.svg",width=6,height=6)
> plot(x,y,col="#ff000018",pch=19,cex=2,main = "Cairo-svg")
> dev.off()
```

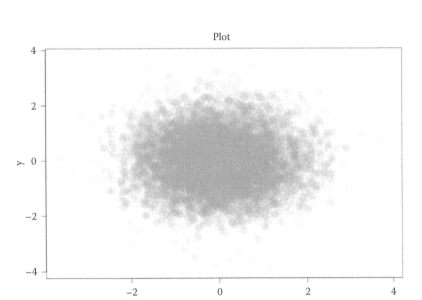

Figure 1.12   Graphic generated by png().

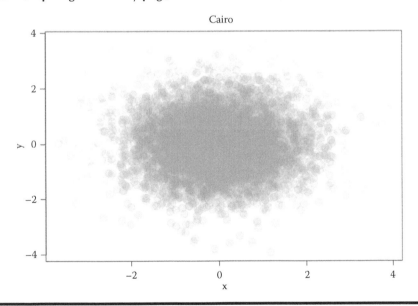

Figure 1.13   Graphic generated by CairoPNG().

Two svg files will be generated in the current directory: plot-svg4.svg and Cairo-svg4.svg. These two files can be displayed in browsers.

1.7.3.2 Three-Dimensional Cross-Sectional View

Next, we'll draw a three-dimensional cross-sectional view and use it to compare the output effect between png() and CairoPNG().

```
> x <- seq(-10, 10, length= 30)
> y <- x
> f <- function(x,y) {r <- sqrt(x^2+y^2); 10 * sin(r)/r}
> z <- outer(x, y, f)
> z[is.na(z)] <- 1

# PNG graphic.
> png(file="plot2.png",width=640,height=480)
> op <- par(bg = "white", mar=c(0,2,3,0)+.1)
> persp(x, y, z, theta = 30, phi = 30, expand = 0.5, col =
"lightblue", ltheta = 120, shade = 0.75, ticktype = "detailed", xlab
= "X", ylab = "Y", zlab = "Sinc(r)", main = "Plot")
> par(op)
> dev.off()

> CairoPNG(file="Cairo2.png",width=640,height=480)
> op <- par(bg = "white", mar=c(0,2,3,0)+.1)
> persp(x, y, z, theta = 30, phi = 30, expand = 0.5, col =
"lightblue", ltheta = 120, shade = 0.75, ticktype = "detailed", xlab
= "X", ylab = "Y", zlab = "Sinc(r)", main = "Cairo")
> par(op)
> dev.off()
```

Two png files will be generated in the current directory: plot2.png and Cairo2.ong, as in Figures 1.14 and 1.15.

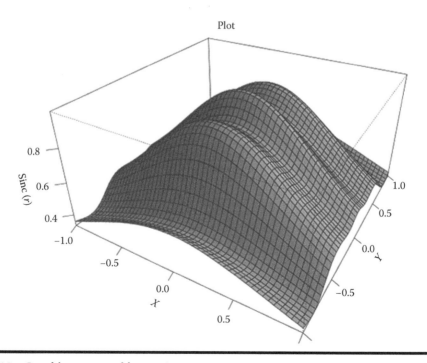

Figure 1.14   Graphic generated by png().

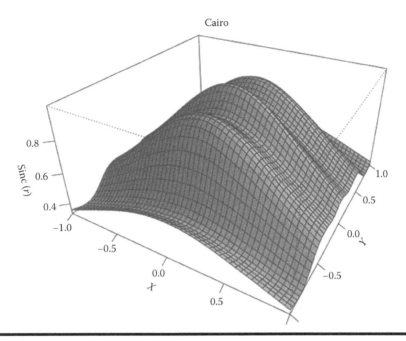

Figure 1.15 Graphic generated by CairoPNG().

1.7.3.3 Graphic with Mass Text in It

Lastly, we compare the output effect of graphics with mass text in it between png() and CairoPNG().

```
# Load MASS.
> library(MASS)

# Load HairEyeColor data set.
> data(HairEyeColor)
> x <- HairEyeColor[,,1]+HairEyeColor[,,2]

> n <- 100
> m <- matrix(sample(c(T,F),n^2,replace=T), nr=n, nc=n)

# PNG Graphic.
> png(file="plot5.png",width=640,height=480)
> biplot(corresp(m, nf=2), main="Plot")
> dev.off()

> CairoPNG(file="Cairo5.png",width=640,height=480)
> biplot(corresp(m, nf=2), main="Cairo")
> dev.off()
```

Two png files will be generated in the current directory: plot5.png and Cairo5.png, as in Figures 1.16 and 1.17.

If we check the properties of these two files we find that a graphic generated by png is 54 kb in size, while that generated by CairoPNG is 43.8 kb in size, which can be seen in Figure 1.18.

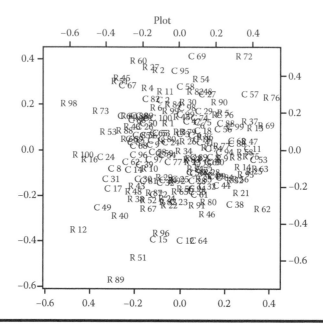

Figure 1.16   Graphic generated by png().

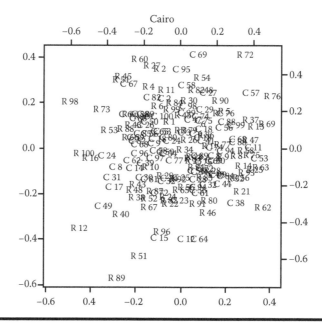

Figure 1.17   Graphic generated by CairoPNG().

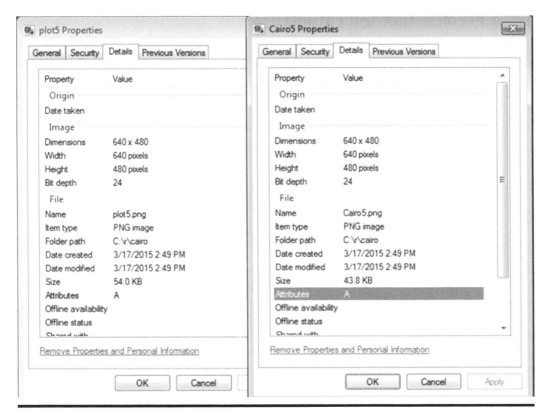

Figure 1.18   Comparison of properties of two files.

The code of graphic output of SVG is

```
> svg(file="plot-svg5.svg",width=6,height=6)
> biplot(corresp(m, nf=2), main="Plot-svg")
> dev.off()

> CairoSVG(file="Cairo-svg5.svg",width=6,height=6)
> biplot(corresp(m, nf=2), main="Cairo-svg")
> dev.off()
```

Two svg files will be generated in the current directory: plot-svg5.svg and Cairo-svg5.svg.

For all three of the cases in the preceding text, it's quite difficult to tell the difference between CairoPNG() and png(). Cairo is just a little lighter and softer. Concerning this issue, Xie Yihui has added that it's hard to tell the difference because now the png() equipment of R adopts Cairo in default. Several years ago png() did not adopt Cairo, so the PNG generated at that time is rather low in quality. In most cases there would be little difference, while sometimes the performance of anti-aliasing will be different.

1.8 A Peculiar Tool Set: caTools

Question

What can R do except statistics?

R is born to be free. It is different from languages like Java and PHP as they are restrained by unified standards. Almost all R packages vary in their naming and syntax, while R is even mixed and matched in its functions. CaTools is such a mixed and matched library, covering several sets of unrelated functions including image processing, encoding and decoding, classifier, vector computing, and scientific computing, which are all convenient in use and powerful in function. It's impossible to describe these packages in a few words. Only the word peculiar can summarizes its features.

1.8.1 Introduction to caTools

caTools is a basic tool package of R, including moving windows statistics, read/write binary images, quick calculation of area under the curve (AUC), LogitBoost classifier, encoder/decoder of base64, and various functions in quick calculation such as round-off, error, sum, and accumulative sum. The following is an introduction of API in caTools.

1. Read/write binary images
   a. read.gif & write.gif: read/write images in GIF format
   b. read.ENVI & write.ENVI: read/write images in ENVI format, such as GIS images
2. Encoder/decoder of Base64
   a. base64encode: encoder
   b. base64decode: decoder
   Note: Base64 is a way to present binary data-based 64 printable characters. As the sixth power of 2 is 64, every 6 bits will be a unit, corresponding to a certain printable character.
3. Quick calculation of area under the curve (AUC)
   a. colAUC: calculate area under the ROC curve (AUC)
   b. combs: find all unordered combination of elements of vector
   c. trapz: trapezoidal rule of numerical integration
4. LogitBoost classifier
   a. LogitBoost: logitBoost classification algorithm
   b. Predict.LogitBoost: prediction based on logitBoost classification algorithm
   c. sample.split: split data into test and train set

5. Quick calculation tools
   a. runmad: calculate median of vector
   b. runmean: calculate mean of vector
   c. runmin & runmax: calculate min and max of vector
   d. runquatile: calculate quantile of vector
   e. runsd: calculate standard deviation of vector
   f. sumexact, cumsumexact: sum without round-off errors, aiming at optimization of accuracy of double data types in programming languages

1.8.2 Installation of caTools

Environment system used in this section:

■ Linux: Ubuntu Server 12.04.2 LTS 64bit
■ R: 3.0.1 x86_64-pc-linux-gnu

Note: caTools supports both Windows 7 and Linux.

We only need one command to install caTools.

```
# Start R.
~ R

# Installation command of caTools.
> install.packages("caTools")
* installing *source* package 'caTools'...
** package 'caTools' successfully unpacked and MD5 sums checked
** libs
g++ -I/usr/share/R/include -DNDEBUG        -fpic  -O3 -pipe  -g  -c
Gif2R.cpp -o Gif2R.o
g++ -I/usr/share/R/include -DNDEBUG        -fpic  -O3 -pipe  -g  -c
GifTools.cpp -o GifTools.o
gcc -std=gnu99 -I/usr/share/R/include -DNDEBUG        -fpic  -O3 -pipe
-g  -c runfunc.c -o runfunc.o
g++ -shared -o caTools.so Gif2R.o GifTools.o runfunc.o -L/usr/lib/R/lib
-lR
installing to/home/conan/R/x86_64-pc-linux-gnu-library/3.0/caTools/
libs
** R
** preparing package for lazy loading
** help
*** installing help indices
** building package indices
** testing if installed package can be loaded
* DONE (caTools)
```

1.8.3 Use of caTools

Load caTools library in R.

```
> library(caTools)
```

1.8.3.1 Read/Write Binary Images: gif

1. Write a gif image.

```
# Extract the data set: datasets::volcano
# Write it into volcano.gif.
> write.gif(volcano, "volcano.gif", col=terrain.colors, flip=TRUE,
scale="always", comment="Maunga Whau Volcano")
```

2. Read a gif image into memory and output it.

```
# Read the image to y.
> y = read.gif("volcano.gif", verbose=TRUE, flip=TRUE)
GIF image header
Global colormap with 256 colors
Comment Extension
Image [61 x 87]: 3585 bytes
GIF Terminator

# Check the attribute of y.
> attributes(y)
$names
[1] "image" "col" "transparent" "comment"

# Check the storage matrix of image.
> class(y$image)
[1] "matrix"

# Number of columns.
> ncol(y$image)
[1] 61

# Number of rows.
> nrow(y$image)
[1] 87

# Extract the first 10 colors from color table.
> head(y$col,10)
 [1] "#00A600FF" "#01A600FF" "#03A700FF" "#04A700FF" "#05A800FF"
"#07A800FF" "#08A900FF"
 [8] "#09A900FF" "#0BAA00FF" "#0CAA00FF"
```

```
# Check the comment of the image.
> y$comment
[1] "Maunga Whau Volcano"

# Draw the image through y, as in Figure 1.19.*
image(y$image, col=y$col, main=y$comment, asp=1)
```

————————————————

* The readers may run the codes and generate color images.

3. Create a Git animation.

```
> x <- y <- seq(-4*pi, 4*pi, len=200)
> r <- sqrt(outer(x^2, y^2, "+"))
> image = array(0, c(200, 200, 10))
> for(i in 1:10) image[,,i] = cos(r-(2*pi*i/10))/(r^.25)
> write.gif(image, "wave.gif", col="rainbow")
> y = read.gif("wave.gif")
> for(i in 1:10) image(y$image[,,i], col=y$col, breaks=(0:256)-0.5,
asp=1)
```

A wave.gif file will be generated in the current directory, as in Figure 1.20. The demonstration effect of animation can be seen in the online sources of this book, file wave.gif.

We can see that caTools has the same function of outputting GIF animation with the animation package made by Xie Yihui. We do not need to rely on other software libraries when we use caTools to make gif animation, but the saveGIF function of animation package needs to rely on third-party software such as ImageMagick and GraphicsMagick.

1.8.3.2 Encoder/Decoder of Base64

Base64 is often used to process text data, to present, transfer, and store some binary data, including e-mail of MIME, e-mail via MIME, and complex data stored in XML. The encoding and

Figure 1.19   Topographic map of volcano.

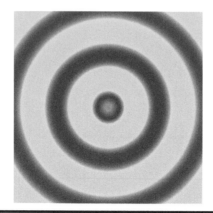

Figure 1.20   Animation of concentric circles.

decoding of base64 supports only a class of vector but not data.frame and list. Here is the code to encode and decode a Boolean vector:

```
# Set the size of each element.
> size=1
> x = (10*runif(10)>5)
> y = base64encode(x, size=size)
> z = base64decode(y, typeof(x), size=size)

# Original Data
> x
 [1] FALSE FALSE FALSE FALSE TRUE TRUE FALSE FALSE FALSE FALSE

# Ciphertext after encoding.
> y
 [1] "AAAAAEBAAAAA=="

# Plaintext after decoding.
> z
 [1] FALSE FALSE FALSE FALSE TRUE TRUE FALSE FALSE FALSE FALSE
```

Here is the code to encode and decode a character string:

```
> x = "Hello R!!" # character
> y = base64encode(x)
> z = base64decode(y, typeof(x))

# Original data.
> x
 [1] "Hello R!!"
```

```
# Ciphertext after encoding.
> y
  [1] "SGVsbG8gUiEh"

# Plaintext after decoding.
> z
  [1] "Hello R!!"
```

Error test: Encode and decode a data frame.

```
> data(iris)
> class(iris)
  [1] "data.frame"

# Original data.
> head(x)
  Sepal.Length Sepal.Width Petal.Length Petal.Width Species
1          5.1         3.5          1.4         0.2  setosa
2          4.9         3.0          1.4         0.2  setosa
3          4.7         3.2          1.3         0.2  setosa
4          4.6         3.1          1.5         0.2  setosa
5          5.0         3.6          1.4         0.2  setosa
6          5.4         3.9          1.7         0.4  setosa

# Encode the data frame.
> base64encode(x)
Error in writeBin(x, raw(), size = size, endian = endian) :
  can only write vector objects
```

1.8.3.3 ROC Curve

ROC curve is the abbreviation for receiver operating characteristic curve, which is an analytical tool of row pictures. It is used mainly in selecting the optimal signal detection model, giving up sub-optimal models and setting the optimal threshold value in a model. The ROC curve was invented by electronics engineers and radar engineers during the Second World War, for signal detection of enemy carriers on the battlefield. It was then introduced to psychology to make perception tests of signals. ROC analysis has been used in medicine, wireless, biology, and criminal psychology for decades. It has also recently gained substantial applications in machine learning and data mining.

Extract data set Mass::cats, the three columns are Sex, Bwt (body weight), and Hwt (heart weight).

```
> library(MASS)
# Load the data set.
> data(cats)
# Print the first 6 lines.
```

```
> head(cats)
  Sex Bwt Hwt
1   F 2.0 7.0
2   F 2.0 7.4
3   F 2.0 9.5
4   F 2.1 7.2
5   F 2.1 7.3
6   F 2.1 7.6

# Calculate the AUC of ROC curve and output image, as in Figure 1.21.
> colAUC(cats[,2:3], cats[,1], plotROC=TRUE)
                Bwt        Hwt
F vs. M 0.8338451 0.759048
```

The standard to judge the quality of classifiers (predicting models) by AUC:

■ AUC = 1 is a perfect classifier. We can always get a perfect prediction no matter what threshold value we set by using such a classifier. A perfect classifier does not exist for prediction on most occasions.

■ 0.5 < AUC < 1 is better than a random prediction. The classifier (model) will have predictive value if the threshold value is set appropriately.

■ AUC = 0.5 is the same as a random prediction (like flipping coins). The model does not have predictive value.

■ AUC < 0.5 is worse than a random prediction. But if we always take the opposite side of this prediction, it's better than a random prediction.

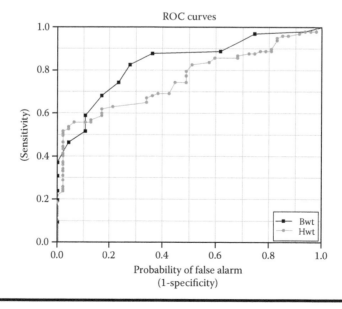

Figure 1.21   ROC curve.

It can be seen from Figure 1.21 that Bwt and Hwt are all within (0.5, 1). So data set cats is a a real and effective data set. If the data set cats is a data set of classification, we can judge the quality of this classifier using AUC on this data set.

1.8.3.4 Unordered Combination of Elements of a Vector

The function combs(v,k) is used to create a matrix of an unordered combination. The columns of the matrix represent the elements in this combination, and the rows of the matrix represent each combination. Parameter v is a vector, and k is numeric, less than the length of v [1:length(v)].

```
> combs(2:5, 3)
     [,1] [,2] [,3]
[1,]    2    3    4
[2,]    2    3    5
[3,]    2    4    5
[4,]    3    4    5

> combs(c("cats", "dogs", "mice"), 2)
      [,1]    [,2]
[1,] "cats"  "dogs"
[2,] "cats"  "mice"
[3,] "dogs"  "mice"

# Create a matrix by quick combination.
> a = combs(1:4, 2)
> a
     [,1] [,2]
[1,]    1    2
[2,]    1    3
[3,]    1    4
[4,]    2    3
[5,]    2    4
[6,]    3    4
> b = matrix(c(1,1,1,2,2,3,2,3,4,3,4,4), 6, 2)
> b
     [,1] [,2]
[1,]    1    2
[2,]    1    3
[3,]    1    4
[4,]    2    3
[5,]    2    4
[6,]    3    4
```

1.8.3.5 Trapezoidal Rule of Numerical Integration

Trapezoidal rule: If you approximate an integrand to a linear function, the integrated part will be an approximation of a trapezoid.

```
# integral of sine function in [0, pi] range suppose to be exactly
2.
# lets calculate it using 10 samples:
> x = (1:10)*pi/10
> trapz(x, sin(x))
[1] 1.934983

> x = (1:1000)*pi/1000
> trapz(x, sin(x))
[1] 1.999993
```

1.8.3.6 LogitBoost Classifier

If you extract the data set datasets::iris, the five columns are Sepan.Length, Sepal.Width, Petal.Length, Petal.Width and Species.

```
> head(iris)
  Sepal.Length Sepal.Width Petal.Length Petal.Width Species
1          5.1         3.5          1.4         0.2 setosa
2          4.9         3.0          1.4         0.2 setosa
3          4.7         3.2          1.3         0.2 setosa
4          4.6         3.1          1.5         0.2 setosa
5          5.0         3.6          1.4         0.2 setosa
6          5.4         3.9          1.7         0.4 setosa

> Data = iris[,-5]
> Label = iris[, 5]

# Train model.
> model = LogitBoost(Data, Label, nIter=20)

# Model data.
> model
$Stump
       feature threshhold sign feature threshhold sign feature threshhold sign
 [1,]        3        1.9   -1       2        2.9   -1       4        1.6    1
 [2,]        4        0.6   -1       3        4.7   -1       3        4.8    1
 [3,]        3        1.9   -1       2        2.0   -1       4        1.7    1
 [4,]        4        0.6   -1       3        1.9    1       3        4.9    1
 [5,]        3        1.9   -1       4        1.6   -1       4        1.3    1
 [6,]        4        0.6   -1       1        6.5    1       2        2.6   -1
 [7,]        3        1.9   -1       3        1.9    1       4        1.7    1
 [8,]        4        0.6   -1       2        2.0   -1       2        3.0   -1
 [9,]        3        1.9   -1       3        5.0   -1       3        5.0    1
[10,]        4        0.6   -1       2        2.9    1       1        4.9   -1
[11,]        3        1.9   -1       3        1.9    1       3        4.4    1
[12,]        4        0.6   -1       2        2.0   -1       4        1.7    1
[13,]        3        1.9   -1       3        5.1   -1       2        3.1   -1
[14,]        4        0.6   -1       2        2.0   -1       3        5.1    1
[15,]        3        1.9   -1       3        1.9    1       1        6.5   -1
[16,]        4        0.6   -1       4        1.6   -1       3        5.1    1
[17,]        3        1.9   -1       2        3.1    1       2        3.1   -1
[18,]        4        0.6   -1       3        1.9    1       1        4.9   -1
```

```
[19,]     3      1.9   -1     2      2.0   -1     4      1.4    1
[20,]     4      0.6   -1     3      5.1   -1     2      2.2   -1


$lablist
[1] setosa     versicolor virginica
Levels: setosa versicolor virginica


attr(,"class")
[1] "LogitBoost"


# Prediction on Classification. Lab only displays the result of classifying,
Prob displays the probability of each classification.
> Lab = predict(model, Data)
> Prob = predict(model, Data, type="raw")


# Merge the result and print out the first 6.
> t = cbind(Lab, Prob)
> head(t)
     Lab setosa  versicolor    virginica
[1,]   1      1  0.017986210 1.522998e-08
[2,]   1      1  0.002472623 3.353501e-04
[3,]   1      1  0.017986210 8.315280e-07
[4,]   1      1  0.002472623 4.539787e-05
[5,]   1      1  0.017986210 1.522998e-08
[6,]   1      1  0.017986210 1.522998e-08
```

For the first six pieces of data, Lab columns show that the data belong to classification 1, setosa. The other three columns—setosa, versicolor, and virginica—represent the probability of that classification. Next we set iterations to compare the result of the classification and that of real data.

```
# Set iterations to 2.
> table(predict(model, Data, nIter=2), Label)
            Label
             setosa versicolor virginica
  setosa         48          0         0
  versicolor      0         45         1
  virginica       0          3        45


# Set iterations to 10.
> table(predict(model, Data, nIter=10), Label)
            Label
             setosa versicolor virginica
  setosa         50          0         0
  versicolor      0         47         0
  virginica       0          1        47


# Default iteration, nIter value of LogitBoost in train.
> table(predict(model, Data), Label)
            Label
             setosa versicolor virginica
  setosa         50          0         0
  versicolor      0         49         0
  virginica       0          0        48
```

From the three preceding tests we can see that the more iteration, the more accurate the model will be. Then we split the train set and test set randomly and make a prediction on the classification.

```
# Extract train set randomly.
> mask = sample.split(Label)

# The 99th record of train set.
> length(which(mask))
[1] 99

# The 51st record of test set.
> length(which(!mask))
[1] 51

# Train model.
> model = LogitBoost(Data[mask,], Label[mask], nIter=10)

# Prediction on classification.
> table(predict(model, Data[!mask,], nIter=2), Label[!mask])
           setosa versicolor virginica
  setosa       16          0         0
  versicolor    0         15         3
  virginica     0          1        12

> table(predict(model, Data[!mask,]), Label[!mask])
           setosa versicolor virginica
  setosa       17          0         0
  versicolor    0         16         4
  virginica     0          1        13
```

1.8.3.7 Quick Calculation Tool: runmean

The moving average is a very popular, simple, and practical analytical index in stock trading. Now we calculate the moving average of time series data. The output is in Figure 1.22.

```
# Take the data set datasets::BJsales.
> BJsales
Time Series:
Start = 1
End = 150
Frequency = 1
  [1] 200.1 199.5 199.4 198.9 199.0 200.2 198.6 200.0 200.3 201.2
201.6 201.5
 [13] 201.5 203.5 204.9 207.1 210.5 210.5 209.8 208.8 209.5 213.2
213.7 215.1
 [25] 218.7 219.8 220.5 223.8 222.8 223.8 221.7 222.3 220.8 219.4
220.1 220.6
 [37] 218.9 217.8 217.7 215.0 215.3 215.9 216.7 216.7 217.7 218.7
222.9 224.9
```

```
 [49] 222.2 220.7 220.0 218.7 217.0 215.9 215.8 214.1 212.3 213.9
214.6 213.6
 [61] 212.1 211.4 213.1 212.9 213.3 211.5 212.3 213.0 211.0 210.7
210.1 211.4
 [73] 210.0 209.7 208.8 208.8 208.8 210.6 211.9 212.8 212.5 214.8
215.3 217.5
 [85] 218.8 220.7 222.2 226.7 228.4 233.2 235.7 237.1 240.6 243.8
245.3 246.0
 [97] 246.3 247.7 247.6 247.8 249.4 249.0 249.9 250.5 251.5 249.0
247.6 248.8
[109] 250.4 250.7 253.0 253.7 255.0 256.2 256.0 257.4 260.4 260.0
261.3 260.4
[121] 261.6 260.8 259.8 259.0 258.9 257.4 257.7 257.9 257.4 257.3
257.6 258.9
[133] 257.8 257.7 257.2 257.5 256.8 257.5 257.0 257.6 257.3 257.5
259.6 261.1
[145] 262.9 263.3 262.8 261.8 262.2 262.7

> plot(BJsales, col="black", main = "Moving Window Means")
> lines(runmean(BJsales, 3), col="red")
> lines(runmean(BJsales, 8), col="green")
> lines(runmean(BJsales,15), col="blue")
> lines(runmean(BJsales,24), col="magenta")
> lines(runmean(BJsales,50), col="cyan")
```

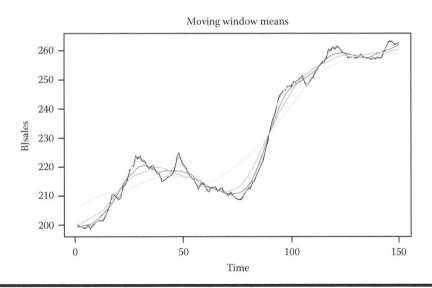

Figure 1.22 Moving average.

There are six lines in Figure 1.22. The black line is the original data, and lines in other colors are moving averages in different units. For example, the red line represents the average of three points, and the green line represents the average of eight points.*

---

* The readers may run the source code to check the color image.

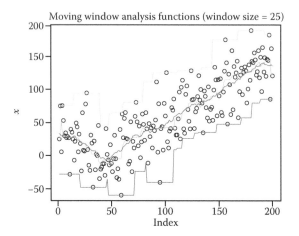

Figure 1.23 Path graph.

1.8.3.8 Combinations of Quick Calculation Tools

Extract maximum path (runmax), minimum path (runmin), mean path (runmean), and median path (runmed) of the data set. The output is in Figure 1.23.

```
> n=200
> x = rnorm(n,sd=30) + abs(seq(n)-n/4)
> plot(x, main = "Moving Window Analysis Functions (window
size=25)")
> lines(runmin (x,k), col="red")
> lines(runmed (x,k), col="green")
> lines(runmean(x,k), col="blue")
> lines(runmax (x,k), col="cyan")
```

1.8.3.9 Exact Summation

There will be errors in all programming languages in making decimal calculations using computers. There is such a problem in R as well. First we do sum by sum().

```
# Summation
> x = c(1, 1e20, 1e40, -1e40, -1e20, -1)
> a = sum(x); print(a)
[1] -1e+20

# Exact Summation
> b = sumexact(x); print(b)
[1] 0
```

If we sum the vector x, the result should be 0, but the result in sum() is −1e+20. This is because there is a calculation error caused by the accuracy of programming. The error no longer exists after correction by sumexact(). Then we do an accumulative sum by cumsum().

```
# Accumulative summation.
> a = cumsum(x); print(a)
[1]   1e+00   1e+20   1e+40   0e+00  -1e+20  -1e+20

# Exact accumulative summation.
> b = cumsumexact(x); print(b)
[1] 1e+00 1e+20 1e+40 1e+20 1e+00 0e+00
```

There is also an error in cumsum(), which needs to be corrected by cumsumexact().

In the end, I'd like to once again use the word peculiar to summarize this tool set. I believe that you've already found its peculiarity.

Chapter 2

Basic Packages of Time Series

This chapter mainly introduces three packages of R that can process time series data. It may help readers to master the data structure and basic use of time series data.

2.1 Basic Time Series Library of R: zoo

Question

How do we process time series data in R?

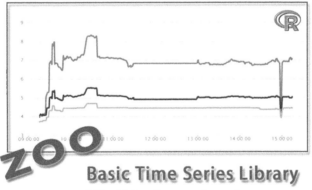

Basic Time Series Library
http://blog.fens.me/r-zoo/

Time series analysis is a statistical method of dynamic data processing. Through time series analysis, we can figure out what is changing in the world! R, as an effective tool for statistical analysis, is very powerful in time series processing. A separate data class defined in R, zoo, is suitable for time series data; it is the basic library for time series and stock analysis. This section introduces the structure of zoo and how to use it in R.

2.1.1 Introduction to zoo

zoo is a class library of R. In the library zoo, an object of class S3 named zoo is defined, which is used to describe regular or irregular time series data. zoo object is an independent object covering index, date, and time, and depends only on the basic R environment. zooreg object extends and behaves like zoo object, but it is used only in analysis of regular time series data. Many other packages of R for time series analysis are based on zoo and zooreg. There are mainly six kind of API in zoo:

1. Basic object
   - zoo: ordered times series object
   - zooreg: regular time series object, extending and behaving like zoo object. It is different from zoo in that it demands that data be continuous
2. Class conversion
   - as.zoo: coercion from and to zoo
   - plot.zoo: plotting method for objects of class zoo
   - xyplot.zoo: plot zoo series with Lattice
   - ggplot2.zoo: plot zoo objects with ggplot2
3. Data operation
   - coredata: extract/replace the core data of zoo
   - index: extract/replace the index of zoo
   - window.zoo: filter data by time
   - merge.zoo: merge two or more zoo objects
   - read.zoo: read zoo sequence from files
   - aggregate.zoo: compute summary statistics of zoo objects
   - rollapply: apply rolling functions
   - rollmean: calculate roll mean of zoo data
4. Processing of NA
   - na.fill: fill NA
   - na.locf: replace NA
   - na.aggregate: replace NA by aggregation
   - na.approx: replace NA using interpolation
   - na.StructTS: replace NA using seasonal Kalman filter
   - na.trim: filter records with NA
5. Auxiliary tools
   - is.regular: check regularity of a series
   - lag.zoo: calculate lag and difference
   - MATCH: value matching
   - ORDER: compute ordering permutation
6. Display control
   - yearqtr: display time in year and quarter
   - yearmon: display time in year and month
   - xblocks: plot contiguous blocks along x-axis
   - make.par.list: format conversion for plot.zoo data and xyplot.zoo data

2.1.2 Installation of zoo

System environment used in this section:

- ■ Windows 7 64bit
- ■ R: 3.0.1 x86_64-w64-mingw32/x64 b4bit

Note: zoo supports both Windows 7 and Linux.
The installation of zoo:

```
# Start R.
~ R

# Install zoo package.
> install.packages("zoo")

# Load zoo package.
> library(zoo)
```

2.1.3 Use of zoo

2.1.3.1 zoo Object

There are two parts in zoo object: core data and index. First is the definition of function:

```
zoo(x = NULL, order.by = index(x), frequency = NULL)
```

X is the core data, which allows class of vector, matrix, and factor. Order.by is the index, which demands uniqueness of fields for ordering. Frequency is the number of observations per unit of time displayed.

The code that follows will create a zoo object indexed by time, as in Figure 2.1. We should pay special attention to the fact that zoo object allows discontinuous time series data.

```
# Define a vector of discontinuous date.
> x.Date <- as.Date("2003-02-01") + c(1, 3, 7, 9, 14) - 1
> x.Date
[1] "2003-02-01" "2003-02-03" "2003-02-07" "2003-02-09" "2003-02-14"
> class(x.Date)
[1] "Date"

# Define a discontinuous zoo object.
> x <- zoo(rnorm(5), x.Date)
> x
2003-02-01 2003-02-03 2003-02-07 2003-02-09 2003-02-14
0.01964254 0.03122887 0.64721059 1.47397924 1.29109889
> class(x)
[1] "zoo"

# Display.
> plot(x)
```

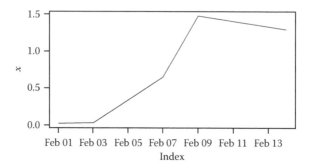

Figure 2.1   Time series graph.

Next, we create some groups of time series data indexed by number. The following code will create a matrix of 4 rows and 3 columns with 12 elements. Create a zoo object y indexed by number 0:10 and output y as in Figure 2.2.

```
> y <- zoo(matrix(1:12, 4, 3),0:10)
> y
0   1 5  9
1   2 6 10
2   3 7 11
3   4 8 12
4   1 5  9
5   2 6 10
6   3 7 11
7   4 8 12
8   1 5  9
9   2 6 10
10  3 7 11

# Each column of the matrix is a time series graph.
> plot(y)
```

2.1.3.2 zooreg Object

First is the definition of functions:

```
zooreg(data, start = 1, end = numeric(), frequency = 1,
deltat = 1, ts.eps = getOption("ts.eps"), order.by = NULL)
```

The following is the parameter description.

■ Data: core data, allows class of vector, matrix, and factor.
■ Start: time part, the time of starting.
■ End: time part, the time of ending.
■ Frequency: number of observations per each time unit displayed.
■ Deltat: sampling period of continuous observation, which doesn't appear with frequency at the same time. For example, the value is 1/12 for data of each month.

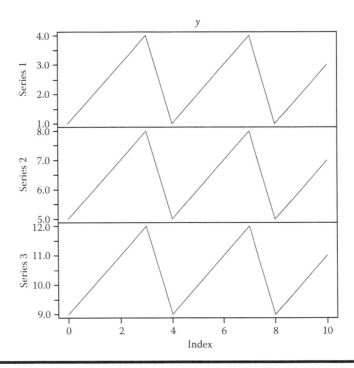

Figure 2.2 Time series graphs in groups.

- Ts.eps: interval of time series. If the interval of data is less than ts.eps, es.pes will be used as the interval. It is set through getOption('ts.eps'), which is 12-05 in default.
- Order.by: index, which demands the uniqueness of fields for ordering. Extend and behave like order.by of zoo.

The following code will create a zooreg object indexed by continuous year (quarter), as in Figure 2.3.

```
> zooreg(1:10, frequency = 4, start = c(1959, 2))
1959(2) 1959(3) 1959(4) 1960(1) 1960(2) 1960(3) 1960(4) 1961(1)
1961(2)
      1       2       3       4       5       6       7       8       9
1961(3)
     10

> as.zoo(ts(1:10, frequency = 4, start = c(1959, 2)))
1959(2) 1959(3) 1959(4) 1960(1) 1960(2) 1960(3) 1960(4) 1961(1)
1961(2)
      1       2       3       4       5       6       7       8       9
1961(3)
     10

> zr<-zooreg(rnorm(10), frequency = 4, start = c(1959, 2))
> plot(zr)
```

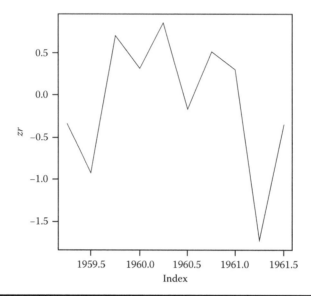

Figure 2.3   Time series graph of zooreg object.

2.1.3.3 *Difference between a zooreg Object and a zoo Object*

The difference between a zoo object and a zooreg object lies in calculating lag and difference.

- Lag: Lag is calculated according to index in zoo, and according to value zooreg.
- Difference: Difference is calculated according to index in zoo, and according to value in zooreg.

Here is an example of a group of discontinuous data (1,2,3,6,7,8):

```
> x <- c(1, 2, 3, 6, 7, 8)
> zz <- zoo(x, x)
> zr <- as.zooreg(zz)

# Lag.
> lag(zz, k = -1)
2 3 6 7 8
1 2 3 6 7

> lag(zr, k = -1)
2 3 4 7 8 9
1 2 3 6 7 8

# Difference.
> diff(zz)
2 3 6 7 8
1 1 3 1 1
```

```
> diff(zr)
2 3 7 8
1 1 1 1
```

2.1.3.4 Class Conversion of zoo Object

First the object is converted from another class to the zoo class.

```
# Convert a vector of other class to class zoo.
> as.zoo(rnorm(5))
          1          2          3          4          5
-0.4892119  0.5740950  0.7128003  0.6282868  1.0289573
> as.zoo(ts(rnorm(5), start = 1981, freq = 12))
   1981(1)    1981(2)    1981(3)    1981(4)    1981(5)
2.3198504  0.5934895 -1.9375893 -1.9888237  1.0944444

# Convert a class ts to class zoo.
> x <- as.zoo(ts(rnorm(5), start = 1981, freq = 12))
> x
1981(1)    1981(2)    1981(3)    1981(4)    1981(5)
1.8822996  1.6436364  0.1260436 -2.0360960 -0.1387474
```

Then the object is converted from the zoo class to the other class.

```
# Convert class zoo to matrix.
> as.matrix(x)
                  x
1981(1)   1.8822996
1981(2)   1.6436364
1981(3)   0.1260436
1981(4)  -2.0360960
1981(5)  -0.1387474

# Convert class zoo to numeric vector.
> as.vector(x)
[1]  1.8822996  1.6436364  0.1260436 -2.0360960 -0.1387474

# Convert class zoo to data frame.
> as.data.frame(x)
                  x
1981(1)   1.8822996
1981(2)   1.6436364
1981(3)   0.1260436
1981(4)  -2.0360960
1981(5)  -0.1387474
```

```
# Convert class zoo to list.
> as.list(x)
[[1]]
   1981(1)    1981(2)    1981(3)    1981(4)    1981(5)
 1.8822996  1.6436364  0.1260436 -2.0360960 -0.1387474
```

2.1.3.5 Draw Time Series Graphs by ggplot2

Because ggplot doesn't support data of the zoo class, we need to call the function zoo::fortify.zoo() by function ggplot2::fortify(). After the conversion of the zoo class to other classes that can be recognized by ggplot2, we can draw graphs of data of the zoo class by ggplot2. The following is the code to draw time series graph of the zoo class using ggplot2, as in Figure 2.4.

```
# Load ggplot2.
> library(ggplot2)
> library(scales)

# Build data object.
> x.Date <- as.Date(paste(2003, 02, c(1, 3, 7, 9, 14), sep = "-"))
> x <- zoo(rnorm(5), x.Date)
> xlow <- x - runif(5)
> xhigh <- x + runif(5)
> z <- cbind(x, xlow, xhigh)

# Display data set.
> z
                    x         xlow        xhigh
2003-02-01 -0.36006612 -0.88751958 0.006247816
2003-02-03  1.35216617  0.97892538 2.076360524
2003-02-07  0.61920828  0.23746410 1.156569424
2003-02-09  0.27516116  0.09978789 0.777878867
2003-02-14  0.02510778 -0.80107410 0.541592929

# Draw a graph using ggplot2.
# Use fortify() to convert data of class zoo to data of data.frame.
> g<-ggplot(aes(x = Index, y = Value), data = fortify(x, melt =
TRUE))
> g<-g+geom_line()
> g<-g+geom_line(aes(x = Index, y = xlow), colour = "red", data =
fortify(xlow))
> g<-g+geom_ribbon(aes(x = Index, y = x, ymin = xlow, ymax = xhigh),
data = fortify(x), fill = "darkgray")
> g<-g+geom_line()
> g<-g+xlab("Index") + ylab("x")
> g
```

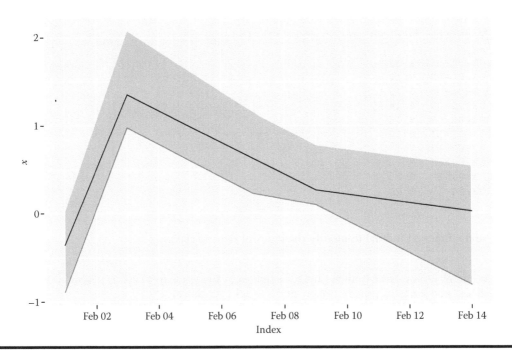

Figure 2.4   Times series graph by ggplot2.

2.1.3.6 Data Operation of zoo Object

Use the function coredata() to modify the core data of the zoo class.

```
> x.date <- as.Date(paste(2003, rep(1:4, 4:1), seq(1,20,2), sep =
"-"))
> x <- zoo(matrix(rnorm(20), ncol = 2), x.date)

# View the core data.
> coredata(x)
             [,1]            [,2]
 [1,] -1.04571765  0.92606273
 [2,] -0.89621126  0.03693769
 [3,]  1.26938716 -1.06620017
 [4,]  0.59384095 -0.23845635
 [5,]  0.77563432  1.49522344
 [6,]  1.55737038  1.17215855
 [7,] -0.36540180 -1.45770721
 [8,]  0.81655645  0.09505623
 [9,] -0.06063478  0.84766496
[10,] -0.50137832 -1.62436453

# Modify the core data.
> coredata(x) <- matrix(1:20, ncol = 2)
```

```
# View the data set after modification.
> x
2003-01-01  1 11
2003-01-03  2 12
2003-01-05  3 13
2003-01-07  4 14
2003-02-09  5 15
2003-02-11  6 16
2003-02-13  7 17
2003-03-15  8 18
2003-03-17  9 19
2003-04-19 10 20
```

Use the function index() to modify the index of the zoo class.

```
> x.date <- as.Date(paste(2003, rep(1:4, 4:1), seq(1,20,2), sep = "-"))
> x <- zoo(matrix(rnorm(20), ncol = 2), x.date)

# View the index.
> index(x)
 [1] "2003-01-01" "2003-01-03" "2003-01-05" "2003-01-07"
"2003-02-09"
 [6] "2003-02-11" "2003-02-13" "2003-03-15" "2003-03-17"
"2003-04-19"

# Modify the index.
> index(x) <- 1:nrow(x)

# View the index after modification.
> index(x)
 [1]  1  2  3  4  5  6  7  8  9 10
```

Use the function window.zoo() to filter data by time.

```
> x.date <- as.Date(paste(2003, rep(1:4, 4:1), seq(1,20,2), sep = "-"))
> x <- zoo(matrix(rnorm(20), ncol = 2), x.date)

# Extract the data dated from 2003-02-01 to 2003-03-01.
> window(x, start = as.Date("2003-02-01"), end = as.Date("2003-03-01"))
2003-02-09  0.7021167 -0.3073809
2003-02-11  2.5071111  0.6210542
2003-02-13 -1.8900271  0.1819022

# Extract data which is dated from 2003-02-01 and has index date in
x.date[1:6].
> window(x, index = x.date[1:6], start = as.Date("2003-02-01"))
2003-02-09 0.7021167 -0.3073809
2003-02-11 2.5071111  0.6210542
```

```
# Extract the data which has index date in x.date[c(4,8,10)].
> window(x, index = x.date[c(4, 8, 10)])
2003-01-07  1.4623515 -1.198597
2003-03-15 -0.5898128  1.318401
2003-04-19 -0.4209979 -1.648222
```

Use merge.zoo() to merge multiple zoo objects.

```
# Create two groups of zoo data.
> y1 <- zoo(matrix(1:10, ncol = 2), 1:5);y1
1 1  6
2 2  7
3 3  8
4 4  9
5 5 10


> y2 <- zoo(matrix(rnorm(10), ncol = 2), 3:7);y2
3  1.4810127  0.13575871
4 -0.3914258  0.06404148
5  0.6018237  1.85017952
6  1.2964150 -0.12927481
7  0.2211769  0.32381709

# Merge the data with same index value.
> merge(y1, y2, all = FALSE)
  y1.1 y1.2       y2.1       y2.2
3    3    8  0.9514985  1.7238941
4    4    9 -1.1131230 -0.2061446
5    5   10  0.6169665 -1.3141951

# Name the customized data columns.
> merge(y1, y2, all = FALSE, suffixes = c("a", "b"))
  a.1 a.2        b.1        b.2
3   3   8  0.9514985  1.7238941
4   4   9 -1.1131230 -0.2061446
5   5  10  0.6169665 -1.3141951

# Merge the data set in a completed one, and fill the blank data
with NA.
> merge(y1, y2, all = TRUE)
  y1.1 y1.2       y2.1       y2.2
1    1    6         NA         NA
2    2    7         NA         NA
3    3    8  0.9514985  1.7238941
4    4    9 -1.1131230 -0.2061446
5    5   10  0.6169665 -1.3141951
6   NA   NA  0.5134937  0.0634741
7   NA   NA  0.3694591 -0.2319775
```

```
# Merge the data set in a completed one, and fill the blank data
with 0.
> merge(y1, y2, all = TRUE, fill = 0)
  y1.1 y1.2       y2.1        y2.2
1    1    6  0.0000000   0.0000000
2    2    7  0.0000000   0.0000000
3    3    8  0.9514985   1.7238941
4    4    9 -1.1131230  -0.2061446
5    5   10  0.6169665  -1.3141951
6    0    0  0.5134937   0.0634741
7    0    0  0.3694591  -0.2319775
```

Use the function aggregate.zoo() to calculate the zoo class.

```
# Create a data set of class zoo, x.
> x.date <- as.Date(paste(2004, rep(1:4, 4:1), seq(1,20,2), sep = "-"))
> x <- zoo(rnorm(12), x.date); x
 2004-01-01  2004-01-03  2004-01-05  2004-01-07  2004-02-09  2004-02-11
 0.67392868  1.95642526 -0.26904101 -1.24455152 -0.39570292  0.09739665
 2004-02-13  2004-03-15  2004-03-17  2004-04-19
-0.23838695 -0.41182796 -1.57721805 -0.79727610

# Create time vector, x.date2.
> x.date2 <- as.Date(paste(2004, rep(1:4, 4:1), 1, sep = "-"));
x.date2
 [1] "2004-01-01" "2004-01-01" "2004-01-01" "2004-01-01" "2004-02-01"
 [6] "2004-02-01" "2004-02-01" "2004-03-01" "2004-03-01" "2004-04-01"

# Calculate mean of x under the time division rule of x.date2.
> x2 <- aggregate(x, x.date2, mean); x2
2004-01-01 2004-02-01 2004-03-01 2004-04-01
 0.2791904 -0.1788977 -0.9945230 -0.7972761
```

2.1.3.7 Functional Processing of Data of zoo Object

Use the function rollapply() to run functional processing on zoo data.

```
> z <- zoo(11:15, as.Date(31:35))

# Calculate the mean of two continuous days  from starting date.
> rollapply(z, 2, mean)
1970-02-01 1970-02-02 1970-02-03 1970-02-04
     11.5       12.5       13.5       14.5

# Calculate the mean of three continuous days from starting date.
> rollapply(z, 3, mean)
1970-02-02 1970-02-03 1970-02-04
        12         13         14
```

Equivalent operation: use rollapply() to implement the operation of aggregate().

```
> z2 <- zoo(rnorm(6))
> rollapply(z2, 3, mean, by = 3) # means of nonoverlapping groups of
3
          2          5
-0.3065197  0.6350963
> aggregate(z2, c(3,3,3,6,6,6), mean) # same
          3          6
-0.3065197  0.6350963
```

Equivalent operation: use rollapply() to implement the operation of rollmean().

```
> rollapply(z2, 3, mean) # uses rollmean which is optimized for mean
          2          3          4          5
-0.3065197 -0.7035811 -0.1672344  0.6350963
> rollmean(z2, 3) # same
          2          3          4          5
-0.3065197 -0.7035811 -0.1672344  0.6350963
```

2.1.3.8 Processing of NA

Use the function na.fill() to fill NA.

```
# Create a zoo object with NA.
> z <- zoo(c(NA, 2, NA, 3, 4, 5, 9, NA))
> z
 1  2  3  4  5  6  7  8
NA  2 NA  3  4  5  9 NA

# Use extend to fill NA, i.e. fill it with the mean of the last and
next item of NA.
> na.fill(z, "extend")
  1   2   3   4   5   6   7   8
2.0 2.0 2.5 3.0 4.0 5.0 9.0 9.0

# Fill NA in a customized method, i.e. fill it with circulation
-(1:3).
> na.fill(z, -(1:3))
 1  2  3  4  5  6  7  8
-1  2 -2  3  4  5  9 -3

# Fill NA with a combination of extend and customized method.
> na.fill(z, c("extend", NA))
 1  2  3  4  5  6  7  8
 2  2 NA  3  4  5  9  9
```

Use the function na.locf() to replace NA.

```
> z <- zoo(c(NA, 2, NA, 3, 4, 5, 9, NA, 11));z
 1  2  3  4  5  6  7  8  9
NA  2 NA  3  4  5  9 NA 11

# Replace NA with the value of its last item.
> na.locf(z)
 2  3  4  5  6  7  8  9
 2  2  3  4  5  9  9 11

# Replace NA with the value of its next item.
> na.locf(z, fromLast = TRUE)
 1  2  3  4  5  6  7  8  9
 2  2  3  3  4  5  9 11 11
```

Use the value of the statistical calculation of function na.aggregate() to replace NA.

```
z <- zoo(c(1, NA, 3:9),c(as.Date("2010-01-01") + 0:2,as.Date("2010-
02-01") + 0:2,as.Date("2011-01-01") + 0:2))
> z
2010-01-01 2010-01-02 2010-01-03 2010-02-01 2010-02-02 2010-02-03
         1         NA          3          4          5          6
2011-01-01 2011-01-02 2011-01-03
         7          8          9

# Replace the NA of the mean of all other items except NA.
> na.aggregate(z)
2010-01-01 2010-01-02 2010-01-03 2010-02-01 2010-02-02 2010-02-03
    1.000      5.375      3.000      4.000      5.000      6.000
2011-01-01 2011-01-02 2011-01-03
    7.000      8.000      9.000

# Replace NA with the mean of groups indexed by year and month.
> na.aggregate(z, as.yearmon)
2010-01-01 2010-01-02 2010-01-03 2010-02-01 2010-02-02 2010-02-03
         1          2          3          4          5          6
2011-01-01 2011-01-02 2011-01-03
         7          8          9

# Replace NA with the mean of groups indexed by month.
> na.aggregate(z, months)
2010-01-01 2010-01-02 2010-01-03 2010-02-01 2010-02-02 2010-02-03
      1.0        5.6        3.0        4.0        5.0        6.0
2011-01-01 2011-01-02 2011-01-03
      7.0        8.0        9.0
```

```
# Replace NA with the mean of groups indexed by year in regular
expression.
> na.aggregate(z, format, "%Y")
2010-01-01 2010-01-02 2010-01-03 2010-02-01 2010-02-02 2010-02-03
       1.0        3.8        3.0        4.0        5.0        6.0
2011-01-01 2011-01-02 2011-01-03
       7.0        8.0        9.0
```

Use the function na.approx() to replace NA with interpolation.

```
> z <- zoo(c(2, NA, 1, 4, 5, 2), c(1, 3, 4, 6, 7, 8));z
 1  3  4  6  7  8
 2 NA  1  4  5  2

> na.approx(z)
        1        3        4        6        7        8
 2.000000 1.333333 1.000000 4.000000 5.000000 2.000000

> na.approx(z, 1:6)
  1   3   4   6   7   8
 2.0 1.5 1.0 4.0 5.0 2.0
```

Use the function na.StructTS() to calculate the seasonal Kalman filter to replace NA, as in Figure 2.5.

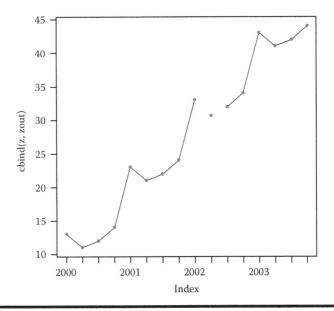

Figure 2.5   Time series to replace NA.

```
z <- zooreg(rep(10 * seq(4), each = 4) + rep(c(3, 1, 2, 4), times = 4),
            start = as.yearqtr(2000), freq = 4)
z[10] <- NA
zout <- na.StructTS(z);zout
plot(cbind(z, zout), screen = 1, col = 1:2, type = c("l", "p"), pch = 20)
```

Use the function na.trim() to remove rows with NA.

```
> xx <- zoo(matrix(c(1, 4, 6, NA, NA, 7), 3), c(2, 4, 6));xx
2 1 NA
4 4 NA
6 6  7

> na.trim(xx)
6 6 7
```

2.1.3.9 Display Format of Data

Output data in the format of "year + quarter."

```
# Output data in the default format of year and quarter.
> x <- as.yearqtr(2000 + seq(0, 7)/4)
> x
[1] "2000 Q1" "2000 Q2" "2000 Q3" "2000 Q4" "2001 Q1" "2001 Q2" "2001 Q3"
[8] "2001 Q4"

# Output data in the customized format of year and quarter.
> format(x, "%Y Quarter%q")
[1] "2000 Quarter 1" "2000 Quarter 2" "2000 Quarter 3" "2000 Quarter 4"
[5] "2001 Quarter 1" "2001 Quarter 2" "2001 Quarter 3" "2001 Quarter 4"

> as.yearqtr("2001 Q2")
[1] "2001 Q2"

> as.yearqtr("2001 q2")
[1] "2001 Q2"

> as.yearqtr("2001-2")
[1] "2001 Q2"
```

Output data in the format of "year + month."

```
# Output in the default format of "year+month"
> x <- as.yearmon(2000 + seq(0, 23)/12)
> x
[1] "January 2000"      "February 2000"  "March 2000"      "April 2000"      "May 2000"
[6] "June 2000"         "July 2000"      "August 2000"     "September 2000"  "October 2000"
[11] "November 2000"    "December 2000"  "January 2001"    "February 2001"   "March 2001"
[16] "April 2000 2001"  "May 2001"       "June 2001"       "July 2001"       "August 2001"
[21] "September 2001"   "October 2001"   "November 2001"   "December 2001"

> as.yearmon("mar07", "%b%y")
[1] NA

> as.yearmon("2007-03-01")
[1] "March 2007"

> as.yearmon("2007-12")
[1] "December 2007"
```

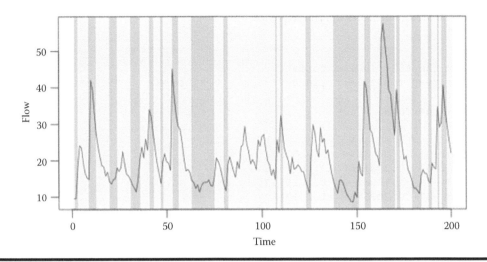

Figure 2.6 Time series graph with divisions.

2.1.3.10 Interval Division

Use the function xblock() to divide the graph in three intervals, (-Inf, 15), [15, 30], (30, Inf), as in Figure 2.6.

```
> set.seed(0)
> flow <- ts(filter(rlnorm(200, mean = 1), 0.8, method = "r"))
> rgb <- hcl(c(0, 0, 260), c = c(100, 0, 100), l = c(50, 90, 50),
alpha = 0.3)
> plot(flow)
> xblocks(flow > 30, col = rgb[1]) ## high values red
> xblocks(flow < 15, col = rgb[3]) ## low value blue
> xblocks(flow >= 15 & flow <= 30, col = rgb[2]) ## the rest gray
```

2.1.3.11 Read Time Series Data from Files and Create zoo Object

First we create a file and name it read.csv:

```
~ vi read.csv

2003-01-01,1.0073644,0.05579711
2003-01-03,-0.2731580,0.06797239
2003-01-05,-1.3096795,-0.20196174
2003-01-07,0.2225738,-1.15801525
2003-02-09,1.1134332,-0.59274327
2003-02-11,0.8373944,0.76606538
2003-02-13,0.3145168,0.03892812
2003-03-15,0.2222181,0.01464681
2003-03-17,-0.8436154,-0.18631697
2003-04-19,0.4438053,1.40059083
```

Read the file and generate the zoo sequence:

```
# Read the file in zoo format.
> r <- read.zoo(file="read.csv",sep = ",", format = "%Y-%m-%d")

# View the data.
> r
                  V2           V3
2003-01-01  1.0073644    0.05579711
2003-01-03 -0.2731580    0.06797239
2003-01-05 -1.3096795   -0.20196174
2003-01-07  0.2225738   -1.15801525
2003-02-09  1.1134332   -0.59274327
2003-02-11  0.8373944    0.76606538
2003-02-13  0.3145168    0.03892812
2003-03-15  0.2222181    0.01464681
2003-03-17 -0.8436154   -0.18631697
2003-04-19  0.4438053    1.40059083

# View the class.
> class(r)
[1] "zoo"
```

We've fully mastered the use of zoo library and zoo object. Now we can start processing time series data in R!

2.2 Extensible Time Series: xts

Question

How do we process complex time series data?

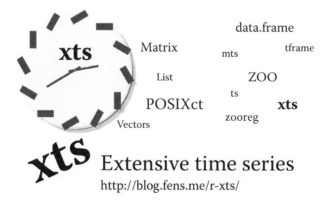

Extensive time series
http://blog.fens.me/r-xts/

This section continues to introduce the extended implementation of zoo. Sometimes time series data may contain complex laws. As a basic library of time series, zoo is designed for universal problems, such as defining stock data and analyzing weather data. But when it comes to other assignments, we may need more auxiliary functions to help fulfill these assignments more efficiently. Xts, as an extension of zoo, provides more functions for data processing and data conversion.

2.2.1 Introduction to xts

Xts, as an extension of time series data (zoo), aims to unify the operation interface of time series. In fact, the xts class extends and behaves like the zoo class and enriches the functions for time series data processing. Its API definition is closer to users, and more practical and simple!

2.2.1.1 Data Structure of xts

The basic extension of xts to zoo is composed of three parts, as in Figure 2.7.

- Index: vector of time/dates
- Core data: based on matrix, and supports all other types that can convert to/from matrix
- Attribute part: attachment information, including format of time zone and index of time/dates

2.2.1.2 Introduction to API of xts

1. Basics of xts
   - xts: define the data type of xts, extending and behaving like zoo
   - coredata.xts: take/replace core data of an xts object
   - xtsAttributes: take and replace noncore xts attributes
   - [.xts]: take subsets of xts objects
   - dimnames.xts: get or set dimnames of an xts object
   - sample_matrix: sample data matrix for xts example and unit testing, simulated 180 observations on 4 variables
   - xtsAPI: xts C API documentation
2. Type conversion
   - as.xts: convert object to and from the xts class
   - as.xts.methods: convert object to and from the xts class
   - plot.xts: plotting methods for xts objects
   - .parseISO8601: output character string (in ISO8601 format) in the POSIXct class, including starting time and ending time of list object
   - firstof: create a time stamp corresponding to the first observation, the POSIXct class
   - lastof: create a time stamp corresponding to the last observation, the POSICXct class
   - indexClass: take and replace the class of an xts object

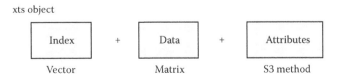

Figure 2.7   Data structure of xts.

- .indexDate: value of date in index
- .indexday: value of date in index, same as.indexDate
- .indexyday: value of year(day) in index
- .indexmday: value of month(day) in index
- .indexwday: value of week(day) in index
- .indexweek: value of week in index
- .indexmon: value of month in index
- .indexyear: value of year in index
- .indexhour: value of hour in index
- .indexmin: value of minute in index
- .indexsec: value of second in index

3. Data processing
   - align.time: align seconds, minutes, and hours to beginning of next period
   - endpoints: take index values by time
   - merge.xts: merge multiple xts objects, rewriting zoo::merge.zoo function
   - rbind.xts: concatenate or bind by row two or more xts objects
   - split_xts: creates a list of xts objects split along time periods
   - na.locf.xts: replace NA value, rewriting zoo::na.locf function

4. Data statistics
   - apply.daily: apply a specified function to each distinct period by day
   - apply.weekly: apply a specified function to each distinct period by week
   - apply.monthly: apply a specified function to each distinct period by month
   - apply.quarterly: apply a specified function to each distinct period by quarter
   - apply.yearly: apply a specified function to each distinct period by year
   - to.period: convert time series data to an OHLC series
   - period.apply: apply function over specified interval
   - period.max: calculate max by period
   - period.min: calculate min by period
   - period.prod: calculate product by period
   - period.sum: calculate sum by period
   - nseconds: number of seconds in data
   - nminutes: number of minutes in data
   - nhours: number of hours in data
   - ndays: number of days in data
   - nweeks: number of weeks in data
   - nmonths: number of months in data
   - nquarters: number of quarters in data
   - nyears: number of years in data
   - periodicity: estimate the periodicity of a time-series-like object

5. Auxiliary tools
   - first: return the first elements of rows of a vector or two-dimensional data object
   - last: return the last elements of rows of a vector or two-dimensional data object
   - timeBased: check if class is time-based
   - timebasedSeq: create a sequence or range of times
   - diff.xts: compute lags and differences of xts objects
   - isOrdered: check if a vector is ordered

- make.index.unique: force time to be unique
- axTicksByTimes: compute *x*-axis tickmark locations by time
- indexTZ: query the time zone of an xts object

2.2.2 Installation of xts

System environment used in this section:

- Windows 7 64bit
- R: 3.0.1 x86_64-w64-mingw32/x64 b4bit

Note: xts supports both Windows 7 and Linux.
Installation of xts is as follows:

```
# Start R.
~ R

# Install xts.
> install.packages("xts")
also installing the dependency 'zoo'

trying URL 'http://mirror.bjtu.edu.cn/cran/bin/windows/contrib/3.0/xts_0.9-7.
zip'
Content type 'application/zip' length 661664 bytes (646 Kb)
opened URL
downloaded 646 Kb

package 'zoo' successfully unpacked and MD5 sums checked
package 'xts' successfully unpacked and MD5 sums checked

# Load xts.
> library(xts)
```

2.2.3 Use of xts

2.2.3.1 Basic Operation of xts Object

View the training data set, sample_matrix, in xts.

```
# Load sample_matrix.
> data(sample_matrix)

# View the first 6 records of sample_matrix.
> head(sample_matrix)
                Open      High       Low     Close
2007-01-02 50.03978 50.11778 49.95041 50.11778
2007-01-03 50.23050 50.42188 50.23050 50.39767
2007-01-04 50.42096 50.42096 50.26414 50.33236
2007-01-05 50.37347 50.37347 50.22103 50.33459
2007-01-06 50.24433 50.24433 50.11121 50.18112
2007-01-07 50.13211 50.21561 49.99185 49.99185
```

Then define an xts object.

```
# Create an xts object
> sample.xts <- as.xts(sample_matrix, descr='my new xts object')

# Xts is an object extending and behaving like zoo.
> class(sample.xts)
[1] "xts" "zoo"

# Print out object structure.
> str(sample.xts)
An 'xts' object on 2007-01-02/2007-06-30 containing:
  Data: num [1:180, 1:4] 50 50.2 50.4 50.4 50.2 ...
 - attr(*, "dimnames")=List of 2
  ..$: NULL
  ..$: chr [1:4] "Open" "High" "Low" "Close"
  Indexed by objects of class: [POSIXct,POSIXt] TZ:
  xts Attributes:
List of 1
 $ descr: chr "my new xts object"

# View the attribute of object descr.
> attr(sample.xts,'descr')
[1] "my new xts object"
```

Query xts data through matching of character string in [].

```
# Pick out data of 2007.
> head(sample.xts['2007'])
               Open      High      Low     Close
2007-01-02 50.03978 50.11778 49.95041 50.11778
2007-01-03 50.23050 50.42188 50.23050 50.39767
2007-01-04 50.42096 50.42096 50.26414 50.33236
2007-01-05 50.37347 50.37347 50.22103 50.33459
2007-01-06 50.24433 50.24433 50.11121 50.18112
2007-01-07 50.13211 50.21561 49.99185 49.99185

# Pick out data of March, 2007.
> head(sample.xts['2007-03/'])
               Open      High      Low     Close
2007-03-01 50.81620 50.81620 50.56451 50.57075
2007-03-02 50.60980 50.72061 50.50808 50.61559
2007-03-03 50.73241 50.73241 50.40929 50.41033
2007-03-04 50.39273 50.40881 50.24922 50.32636
2007-03-05 50.26501 50.34050 50.26501 50.29567
2007-03-06 50.27464 50.32019 50.16380 50.16380

# Pick out data from 2007-03-06 to the end of 2007.
> head(sample.xts['2007-03-06/2007'])
               Open      High      Low     Close
```

```
2007-03-06 50.27464 50.32019 50.16380 50.16380
2007-03-07 50.14458 50.20278 49.91381 49.91381
2007-03-08 49.93149 50.00364 49.84893 49.91839
2007-03-09 49.92377 49.92377 49.74242 49.80712
2007-03-10 49.79370 49.88984 49.70385 49.88698
2007-03-11 49.83062 49.88295 49.76031 49.78806

# Pick out data of 2007-01-03.
> sample.xts['2007-01-03']
            Open    High     Low    Close
2007-01-03 50.2305 50.42188 50.2305 50.39767
```

2.2.3.2 Draw Graphics Using xts Objects

We can draw graphs (Figure 2.8) and K-line chart (Figure 2.9) using xts objects. The following is the code of drawing graph:

```
> data(sample_matrix)
> plot(as.xts(sample_matrix))
Warning message:
In plot.xts(as.xts(sample_matrix)) :
  only the univariate series will be plotted
```

Warning message prompt indicates that only univariate sequence can be drawn. Thus only the first row, sample_matrix[,1], is drawn.

K-line chart:

```
> plot(as.xts(sample_matrix), type='candles')
```

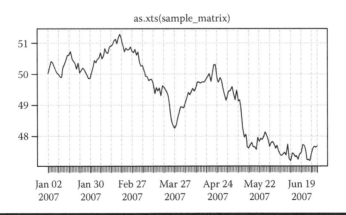

Figure 2.8   Graph drawn using xts object.

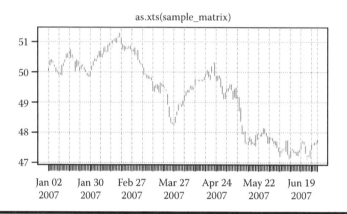

Figure 2.9   K-line chart drawn using xts object.

2.2.3.3 Class Conversion of xts Object

Create firstof() and lastof():

```
# First day of 2000. Omit hour, minute and second.
> firstof(2000)
[1] "2000-01-01 CST"

> firstof(2005,01,01)
[1] "2005-01-01 CST"

# Last second of last day of 2000.
> lastof(2007)
[1] "2007-12-31 23:59:59.99998 CST"

> lastof(2007,10)
[1] "2007-10-31 23:59:59.99998 CST"
```

Create the first observation and the last observation of a time period:

```
# Create first observation and last observation of time of 2000 in
ISO8601.
> .parseISO8601('2000')
$first.time
[1] "2000-01-01 CST"

$last.time
[1] "2000-12-31 23:59:59.99998 CST"

# Create a time period from 2000-05 to 2001-02 in ISO8601.
> .parseISO8601('2000-05/2001-02')
$first.time
[1] "2000-05-01 CST"
```

```
$last.time
[1] "2001-02-28 23:59:59.99998 CST"

> .parseISO8601('2000-01/02')
$first.time
[1] "2000-01-01 CST"

$last.time
[1] "2000-02-29 23:59:59.99998 CST"

> .parseISO8601('T08:30/T15:00')
$first.time
[1] "1970-01-01 08:30:00 CST"

$last.time
[1] "1970-12-31 15:00:59.99999 CST"
```

Create an xts object indexed by time.

```
# Create a time period of class POSIXct.
> x <- timeBasedSeq('2010-01-01/2010-01-02 12:00')
> head(x)
[1] "2010-01-01 00:00:00 CST"
[2] "2010-01-01 00:01:00 CST"
[3] "2010-01-01 00:02:00 CST"
[4] "2010-01-01 00:03:00 CST"
[5] "2010-01-01 00:04:00 CST"
[6] "2010-01-01 00:05:00 CST"
> class(x)
[1] "POSIXt" "POSIXct"

# Create an xts object indexed by time.
> x <- xts(1:length(x), x)
> head(x)
                    [,1]
2010-01-01 00:00:00  1
2010-01-01 00:01:00  2
2010-01-01 00:02:00  3
2010-01-01 00:03:00  4
2010-01-01 00:04:00  5
2010-01-01 00:05:00  6
> indexClass(x)
[1] "POSIXt" "POSIXct"
```

Format the display of time index.

```
# Format the display of time index by regularization.
> indexFormat(x) <- "%Y-%b-%d%H:%M:%OS3"
```

```
> head(x)
                            [,1]
2010-January-01 00:00:00.000    1
2010-January-01 00:01:00.000    2
2010-January-01 00:02:00.000    3
2010-January-01 00:03:00.000    4
2010-January-01 00:04:00.000    5
2010-January-01 00:05:00.000    6
```

Check the time index.

```
# Take the time index by hour.
> .indexhour(head(x))
[1] 0 0 0 0 0 0

# Take the time index by minute.
> .indexmin(head(x))
[1] 0 1 2 3 4 5
```

2.2.3.4 Data Processing of xts Objects

Align data.

```
> x <- Sys.time() + 1:30
# Align data every 10 seconds. The field of second is an integral
multiple of 10.
> align.time(x, 10)
 [1] "2013-11-18 15:42:30 CST" "2013-11-18 15:42:30 CST"
 [3] "2013-11-18 15:42:30 CST" "2013-11-18 15:42:40 CST"
 [5] "2013-11-18 15:42:40 CST" "2013-11-18 15:42:40 CST"
 [7] "2013-11-18 15:42:40 CST" "2013-11-18 15:42:40 CST"
 [9] "2013-11-18 15:42:40 CST" "2013-11-18 15:42:40 CST"
[11] "2013-11-18 15:42:40 CST" "2013-11-18 15:42:40 CST"
[13] "2013-11-18 15:42:40 CST" "2013-11-18 15:42:50 CST"
[15] "2013-11-18 15:42:50 CST" "2013-11-18 15:42:50 CST"
[17] "2013-11-18 15:42:50 CST" "2013-11-18 15:42:50 CST"
[19] "2013-11-18 15:42:50 CST" "2013-11-18 15:42:50 CST"
[21] "2013-11-18 15:42:50 CST" "2013-11-18 15:42:50 CST"
[23] "2013-11-18 15:42:50 CST" "2013-11-18 15:43:00 CST"
[25] "2013-11-18 15:43:00 CST" "2013-11-18 15:43:00 CST"
[27] "2013-11-18 15:43:00 CST" "2013-11-18 15:43:00 CST"
[29] "2013-11-18 15:43:00 CST" "2013-11-18 15:43:00 CST"

# Align data every 60 seconds. The field of second is 0 while the
field of minute is integer.
> align.time(x, 60)
 [1] "2013-11-18 15:43:00 CST" "2013-11-18 15:43:00 CST"
 [3] "2013-11-18 15:43:00 CST" "2013-11-18 15:43:00 CST"
```

```
 [5]  "2013-11-18 15:43:00 CST" "2013-11-18 15:43:00 CST"
 [7]  "2013-11-18 15:43:00 CST" "2013-11-18 15:43:00 CST"
 [9]  "2013-11-18 15:43:00 CST" "2013-11-18 15:43:00 CST"
[11]  "2013-11-18 15:43:00 CST" "2013-11-18 15:43:00 CST"
[13]  "2013-11-18 15:43:00 CST" "2013-11-18 15:43:00 CST"
[15]  "2013-11-18 15:43:00 CST" "2013-11-18 15:43:00 CST"
[17]  "2013-11-18 15:43:00 CST" "2013-11-18 15:43:00 CST"
[19]  "2013-11-18 15:43:00 CST" "2013-11-18 15:43:00 CST"
[21]  "2013-11-18 15:43:00 CST" "2013-11-18 15:43:00 CST"
[23]  "2013-11-18 15:43:00 CST" "2013-11-18 15:43:00 CST"
[25]  "2013-11-18 15:43:00 CST" "2013-11-18 15:43:00 CST"
[27]  "2013-11-18 15:43:00 CST" "2013-11-18 15:43:00 CST"
[29]  "2013-11-18 15:43:00 CST" "2013-11-18 15:43:00 CST"
```

Divide data by time and calculate.

```
> xts.ts <- xts(rnorm(231),as.Date(13514:13744,origin="1970-01-01"))

# Calculate mean by month and display the result by the last day of
each month.
> apply.monthly(xts.ts,mean)
                   [,1]
2007-01-31  0.17699984
2007-02-28  0.30734220
2007-03-31 -0.08757189
2007-04-30  0.18734688
2007-05-31  0.04496954
2007-06-30  0.06884836
2007-07-31  0.25081814
2007-08-19 -0.28845938

# Calculate the customized function (variance) and display the
result by the last day of each month.
> apply.monthly(xts.ts,function(x) var(x))
                  [,1]
2007-01-31 0.9533217
2007-02-28 0.9158947
2007-03-31 1.2821450
2007-04-30 1.2805976
2007-05-31 0.9725438
2007-06-30 1.5228904
2007-07-31 0.8737030
2007-08-19 0.8490521

# Calculate mean by quarter and display the result by the last day
of each quarter.
> apply.quarterly(xts.ts,mean)
                  [,1]
2007-03-31 0.12642053
2007-06-30 0.09977926
2007-08-19 0.04589268
```

```
# Calculate mean by year and display the result by the last day of
each year.
> apply.yearly(xts.ts,mean)
                  [,1]
2007-08-19 0.09849522
```

Use to.period() to divide data by interval.

```
> data(sample_matrix)
# Divide the matrix data by month.
> to.period(sample_matrix)
           sample_matrix.Open sample_matrix.High sample_matrix.Low sample_
matrix.Close
2007-01-31        50.03978           50.77336           49.76308          50.22578
2007-02-28        50.22448           51.32342           50.19101          50.77091
2007-03-31        50.81620           50.81620           48.23648          48.97490
2007-04-30        48.94407           50.33781           48.80962          49.33974
2007-05-31        49.34572           49.69097           47.51796          47.73780
2007-06-30        47.74432           47.94127           47.09144          47.76719
> class(to.period(sample_matrix))
[1] "matrix"

# Divide data of class xts by month.
> samplexts <- as.xts(sample_matrix)
> to.period(samplexts)
           samplexts.Open samplexts.High samplexts.Low samplexts.Close
2007-01-31       50.03978       50.77336       49.76308        50.22578
2007-02-28       50.22448       51.32342       50.19101        50.77091
2007-03-31       50.81620       50.81620       48.23648        48.97490
2007-04-30       48.94407       50.33781       48.80962        49.33974
2007-05-31       49.34572       49.69097       47.51796        47.73780
2007-06-30       47.74432       47.94127       47.09144        47.76719
> class(to.period(samplexts))
[1] "xts" "zoo"
```

Use endpoints() to divide the index data by interval.

```
> data(sample_matrix)

# Divide by month.
> endpoints(sample_matrix)
[1]    0  30  58  89 119 150 180

# Divide by every 7 days.
> endpoints(sample_matrix, 'days',k=7)
 [1]    0   6  13  20  27  34  41  48  55  62  69  76  83  90  97 104
111 118 125
[20] 132 139 146 153 160 167 174 180

# Divide by week.
> endpoints(sample_matrix, 'weeks')
 [1]    0   7  14  21  28  35  42  49  56  63  70  77  84  91  98 105
112 119 126
[20] 133 140 147 154 161 168 175 180
```

```
# Divide by month.
> endpoints(sample_matrix, 'months')
[1]    0   30   58   89  119  150  180
```

Use merge() to merge data by column.

```
# Create 2 data sets of xts.
> (x <- xts(4:10, Sys.Date()+4:10))
           [,1]
2013-11-22    4
2013-11-23    5
2013-11-24    6
2013-11-25    7
2013-11-26    8
2013-11-27    9
2013-11-28   10
> (y <- xts(1:6, Sys.Date()+1:6))
           [,1]
2013-11-19    1
2013-11-20    2
2013-11-21    3
2013-11-22    4
2013-11-23    5
2013-11-24    6

# Merge data by columns and fill NA with the blanks.
> merge(x,y)
            x   y
2013-11-19 NA   1
2013-11-20 NA   2
2013-11-21 NA   3
2013-11-22  4   4
2013-11-23  5   5
2013-11-24  6   6
2013-11-25  7  NA
2013-11-26  8  NA
2013-11-27  9  NA
2013-11-28 10  NA

# Merge data by index.
> merge(x,y, join='inner')
           x y
2013-11-22 4 4
2013-11-23 5 5
2013-11-24 6 6

# Merge data by the left side.
> merge(x,y, join='left')
           x   y
2013-11-22  4   4
2013-11-23  5   5
```

```
2013-11-24  6  6
2013-11-25  7 NA
2013-11-26  8 NA
2013-11-27  9 NA
2013-11-28 10 NA
```

Use rbind() to merge data by row.

```
> x <- xts(1:3, Sys.Date()+1:3)

# Merge data by row.
> rbind(x,x)
            [,1]
2013-11-19   1
2013-11-19   1
2013-11-20   2
2013-11-20   2
2013-11-21   3
2013-11-21   3
```

Use split() to split data by row.

```
> data(sample_matrix)
> x <- as.xts(sample_matrix)

# Split by month, and print out the data of first month.
> split(x)[[1]]
               Open      High       Low     Close
2007-01-02 50.03978 50.11778 49.95041 50.11778
2007-01-03 50.23050 50.42188 50.23050 50.39767
2007-01-04 50.42096 50.42096 50.26414 50.33236
2007-01-05 50.37347 50.37347 50.22103 50.33459
2007-01-06 50.24433 50.24433 50.11121 50.18112
2007-01-07 50.13211 50.21561 49.99185 49.99185
2007-01-08 50.03555 50.10363 49.96971 49.98806
2007-01-09 49.99489 49.99489 49.80454 49.91333
2007-01-10 49.91228 50.13053 49.91228 49.97246
2007-01-11 49.88529 50.23910 49.88529 50.23910
2007-01-12 50.21258 50.35980 50.17176 50.28519
2007-01-13 50.32385 50.48000 50.32385 50.41286
2007-01-14 50.46359 50.62395 50.46359 50.60145
2007-01-15 50.61724 50.68583 50.47359 50.48912
2007-01-16 50.62024 50.73731 50.56627 50.67835
2007-01-17 50.74150 50.77336 50.44932 50.48644
2007-01-18 50.48051 50.60712 50.40269 50.57632
2007-01-19 50.41381 50.55627 50.41278 50.41278
2007-01-20 50.35323 50.35323 50.02142 50.02142
2007-01-21 50.16188 50.42090 50.16044 50.42090
2007-01-22 50.36008 50.43875 50.21129 50.21129
2007-01-23 50.03966 50.16961 50.03670 50.16961
```

```
2007-01-24 50.10953 50.26942 50.06387 50.23145
2007-01-25 50.20738 50.28268 50.12913 50.24334
2007-01-26 50.16008 50.16008 49.94052 50.07024
2007-01-27 50.06041 50.09777 49.97267 50.01091
2007-01-28 49.96586 50.00217 49.87468 49.88096
2007-01-29 49.85624 49.93038 49.76308 49.91875
2007-01-30 49.85477 50.02180 49.77242 50.02180
2007-01-31 50.07049 50.22578 50.07049 50.22578

# Split by week, and print out the data of first two weeks.
> split(x, f="weeks")[[1]]
              Open      High       Low    Close
2007-01-02 50.03978 50.11778 49.95041 50.11778
2007-01-03 50.23050 50.42188 50.23050 50.39767
2007-01-04 50.42096 50.42096 50.26414 50.33236
2007-01-05 50.37347 50.37347 50.22103 50.33459
2007-01-06 50.24433 50.24433 50.11121 50.18112
2007-01-07 50.13211 50.21561 49.99185 49.99185
2007-01-08 50.03555 50.10363 49.96971 49.98806
> split(x, f="weeks")[[2]]
              Open      High       Low    Close
2007-01-09 49.99489 49.99489 49.80454 49.91333
2007-01-10 49.91228 50.13053 49.91228 49.97246
2007-01-11 49.88529 50.23910 49.88529 50.23910
2007-01-12 50.21258 50.35980 50.17176 50.28519
2007-01-13 50.32385 50.48000 50.32385 50.41286
2007-01-14 50.46359 50.62395 50.46359 50.60145
2007-01-15 50.61724 50.68583 50.47359 50.48912
```

Processing of NA.

```
x <- xts(1:10, Sys.Date()+1:10)
> x[c(1,2,5,9,10)] <- NA
> x
            [,1]
2013-11-19   NA
2013-11-20   NA
2013-11-21    3
2013-11-22    4
2013-11-23   NA
2013-11-24    6
2013-11-25    7
2013-11-26    8
2013-11-27   NA
2013-11-28   NA

# Replace NA with the prior observation of NA.
> na.locf(x)
            [,1]
2013-11-19   NA
2013-11-20   NA
2013-11-21    3
```

```
2013-11-22    4
2013-11-23    4
2013-11-24    6
2013-11-25    7
2013-11-26    8
2013-11-27    8
2013-11-28    8

# Replace NA with the next observation of NA.
> na.locf(x, fromLast=TRUE)
             [,1]
2013-11-19    3
2013-11-20    3
2013-11-21    3
2013-11-22    4
2013-11-23    6
2013-11-24    6
2013-11-25    7
2013-11-26    8
2013-11-27    NA
2013-11-28    NA
```

2.2.3.5 Statistical Calculation of xts Objects

We can run statistical calculation on xts objects, like take starting time and ending time, calculate time interval, and calculate statistical indicators by time period.

1. Take the starting time and ending time of xts objects:

```
> xts.ts <- xts(rnorm(231),as.Date(13514:13744,origin="1970-01-01"))

# Take starting time.
> start(xts.ts)
[1] "2007-01-01"

# Take ending time.
> end(xts.ts)
[1] "2007-08-19"

# Print out starting time and ending time in the unit of day.
> periodicity(xts.ts)
Daily periodicity from 2007-01-01 to 2007-08-19
```

2. Calculate time interval:

```
> data(sample_matrix)

# Calculate the number of days.
> ndays(sample_matrix)
[1] 180
```

```
# Calculate the number of weeks.
> nweeks(sample_matrix)
[1] 26

# Calculate the number of months.
> nmonths(sample_matrix)
[1] 6

# Calculate the number of quarters.
> nquarters(sample_matrix)
[1] 2

# Calculate the number of years.
> nyears(sample_matrix)
[1] 1
```

3. Calculate statistical indicators by time period:

```
> zoo.data <- zoo(rnorm(31)+10,as.Date(13514:13744,ori
gin="1970-01-01"))

# Take index of time period by week.
> ep <- endpoints(zoo.data,'weeks')
> ep
 [1]   0   7  14  21  28  35  42  49  56  63  70  77  84  91  98 105
112 119
[19] 126 133 140 147 154 161 168 175 182 189 196 203 210 217 224 231

# Calculate mean by week.
> period.apply(zoo.data, INDEX=ep, FUN=function(x) mean(x))
2007-01-07 2007-01-14 2007-01-21 2007-01-28 2007-02-04 2007-02-11
 10.200488   9.649387  10.304151   9.864847  10.382943   9.660175
2007-02-18 2007-02-25 2007-03-04 2007-03-11 2007-03-18 2007-03-25
  9.857894  10.495037   9.569531  10.292899   9.651616  10.089103
2007-04-01 2007-04-08 2007-04-15 2007-04-22 2007-04-29 2007-05-06
  9.961048  10.304860   9.658432   9.887531  10.608082   9.747787
2007-05-13 2007-05-20 2007-05-27 2007-06-03 2007-06-10 2007-06-17
 10.052955   9.625730  10.430030   9.814703  10.224869   9.509881
2007-06-24 2007-07-01 2007-07-08 2007-07-15 2007-07-22 2007-07-29
 10.187905  10.229310  10.261725   9.855776   9.445072  10.482020
2007-08-05 2007-08-12 2007-08-19
  9.844531  10.200488   9.649387

# Calculate max by week.
> head(period.max(zoo.data, INDEX=ep))
                [,1]
2007-01-07 12.05912
2007-01-14 10.79286
2007-01-21 11.60658
2007-01-28 11.63455
2007-02-04 12.05912
2007-02-11 10.67887
```

```
# Calculate min by week.
> head(period.min(zoo.data, INDEX=ep))
                [,1]
2007-01-07 8.874509
2007-01-14 8.534655
2007-01-21 9.069773
2007-01-28 8.461555
2007-02-04 9.421085
2007-02-11 8.534655

# Calculate the index value by week.
> head(period.prod(zoo.data, INDEX=ep))
                [,1]
2007-01-07 11140398
2007-01-14  7582350
2007-01-21 11930334
2007-01-28  8658933
2007-02-04 12702505
2007-02-11  7702767
```

2.2.3.6 Time Series Operations of xts Objects

Check the class of time.

```
# Sys.time() is class POSIXct.
> class(Sys.time());timeBased(Sys.time())
[1] "POSIXct" "POSIXt"
[1] TRUE

# Sys.Date() is class Date.
> class(Sys.Date());timeBased(Sys.Date())
[1] "Date"
[1] TRUE

# 20070101 is not a class of time.
> class(20070101);timeBased(20070101)
[1] "numeric"
[1] FALSE
```

Use timeBasedSeq() to create time series.

```
# By year.
> timeBasedSeq('1999/2008')
 [1] "1999-01-01" "2000-01-01" "2001-01-01" "2002-01-01"
"2003-01-01"
 [6] "2004-01-01" "2005-01-01" "2006-01-01" "2007-01-01"
"2008-01-01"
```

```
# By month.
> head(timeBasedSeq('199901/2008'))
[1] "December 1998" "January 1999"   "February 1999"   "March
1999"    "April 1999"
[6] "May 1999"

# By day.
> head(timeBasedSeq('199901/2008/d'),40)
 [1] "December 1998" "January 1999"   "January 1999"   "January
1999"    "January 1999"
 [6] "January 1999"   "January 1999"   "January 1999"   "January
1999"    "January 1999"
[11] "January 1999"   "January 1999"   "January 1999"   "January
1999"    "January 1999"
[16] "January 1999"   "January 1999"   "January 1999"   "January
1999"    "January 1999"
[21] "January 1999"   "January 1999"   "January 1999"   "January
1999"    "January 1999"
[26] "January 1999"   "January 1999"   "January 1999"   "January
1999"    "January 1999"
[31] "January 1999"   "January 1999"   "February 1999"   "February
1999"    "February 1999"
[36] "February 1999"   "February 1999"   "February 1999"   "February
1999"    "February 1999"

#  Create a data set of 100 minutes by quantity.
> timeBasedSeq('20080101 0830',length=100)
$from
[1] "2008-01-01 08:30:00 CST"
$to
[1] NA
$by
[1] "mins"
$length.out
[1] 100
```

Take data by index using first() and last()

```
> x <- xts(1:100, Sys.Date()+1:100)
> head(x)
           [,1]
2013-11-19    1
2013-11-20    2
2013-11-21    3
2013-11-22    4
2013-11-23    5
2013-11-24    6

# Take the first 10 pieces of data.
> first(x, 10)
           [,1]
```

```
2013-11-19    1
2013-11-20    2
2013-11-21    3
2013-11-22    4
2013-11-23    5
2013-11-24    6
2013-11-25    7
2013-11-26    8
h
2013-11-27    9
2013-11-28    10

# Take the data of the first day.
> first(x, '1 day')
           [,1]
2013-11-19    1

# Take the data of the last week.
> last(x, '1 weeks')
           [,1]
2014-02-24   98
2014-02-25   99
2014-02-26  100
```

Calculate lags and differences by lag() and diff().

```
> x <- xts(1:5, Sys.Date()+1:5)

# Set the lag to 1.
> lag(x)
           [,1]
2013-11-19   NA
2013-11-20    1
2013-11-21    2
2013-11-22    3
2013-11-23    4

# Set lag to -1 and remove NA.
> lag(x, k=-1, na.pad=FALSE)
           [,1]
2013-11-19    2
2013-11-20    3
2013-11-21    4
2013-11-22    5

# First order difference.
> diff(x)
           [,1]
2013-11-19   NA
2013-11-20    1
```

```
2013-11-21    1
2013-11-22    1
2013-11-23    1

# Second order difference.
> diff(x, lag=2)
            [,1]
2013-11-19    NA
2013-11-20    NA
2013-11-21     2
2013-11-22     2
2013-11-23     2
```

Use isOrdered() to check if the vector is ordered.

```
> isOrdered(1:10, increasing=TRUE)
[1] TRUE

> isOrdered(1:10, increasing=FALSE)
[1] FALSE

> isOrdered(c(1,1:10), increasing=TRUE)
[1] FALSE

> isOrdered(c(1,1:10), increasing=TRUE, strictly=FALSE)
[1] TRUE
```

Use make.index.unique() to force unique index.

```
> x <- xts(1:5, as.POSIXct("2011-01-21") + c(1,1,1,2,3)/1e3)
> x
                          [,1]
2011-01-21 00:00:00.000    1
2011-01-21 00:00:00.000    2
2011-01-21 00:00:00.000    3
2011-01-21 00:00:00.002    4
2011-01-21 00:00:00.003    5

# Guarantee the uniqueness of index by adding millisecond precision.
> make.index.unique(x)
                             [,1]
2011-01-21 00:00:00.000999    1
2011-01-21 00:00:00.001000    2
2011-01-21 00:00:00.001001    3
2011-01-21 00:00:00.002000    4
2011-01-21 00:00:00.003000    5
```

Query time zone of xts objects.

```
> x <- xts(1:10, Sys.Date()+1:10)

# Query time zone.
> indexTZ(x)
[1] "UTC"
> tzone(x)
[1] "UTC"

> str(x)
An 'xts' object on 2013-11-19/2013-11-28 containing:
  Data: int [1:10, 1] 1 2 3 4 5 6 7 8 9 10
 Indexed by objects of class: [Date] TZ: UTC
  xts Attributes:
NULL
```

xts provides more API support than time series of the zoo class. Thus we possess a more convenient tool to make conversion and deformation of time series data.

2.3 Visualization of Time Series: plot.xts

Question

How do we visualize time series data?

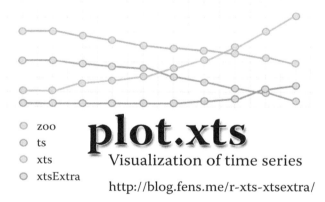

○ zoo
○ ts
○ xts
○ xtsExtra

plot.xts
Visualization of time series
http://blog.fens.me/r-xts-xtsextra/

A blog post of r-bloggers, *plot.xts is wonderful!*, gave me great motivation to continue to explore the power of xts (http://www.r-bloggers.com/plot-xts-is-wonderful/). xts extends the basic data structure of zoo, and provides richer functions. xtsExtra library provides a simple but effective graphic function plot.xts from the perspective of visualization. This section shows how to visualize time-series-like xts objects by plot.xts.

2.3.1 Introduction to xtsExtra

xtsExtra is a supplementary package of xts. xtsExtra was first published in Google Summer of Code 2012. Plot.xts() is a main function provided by xtsExtra. There is a difference between xts:plot.xts() and xtsExtra::plot.xts(), which is discussed in this section.

2.3.2 Installation of xtsExtra

System environment used in this section:

- ▪ Windows 7 64bit
- ▪ R: 3.0.1 x86_64-w64-mingw32/x64 b4bit

Note: xtsExtra supports both Windows 7 and Linux. Since xtsExtra isn't published in CRAN, we need to download it in R-Forge.

```
# Start R.
~ R

# Download xtsExtra from R-Forge.
>install.packages("xtsExtra",repos="http://R-Forge.R-project.org")

# Load xtsExtra.
> library(xtsExtra)
```

XtsExtra::plot.xts() has covered xts::plot.xts().

2.3.3 Use of xtsExtra

The parameter list of plot.xts() is as follows:

```
> names(formals(plot.xts))
 [1] "x"            "y"           "screens"       "layout.screens"
"..."
 [6] "yax.loc"     "auto.grid"   "major.ticks"   "minor.ticks"
"major.format"
[11] "bar.col.up"  "bar.col.dn"  "candle.col"   "xy.labels"    "xy.
lines"
[16] "ylim"        "panel"       "auto.legend"   "legend.names"
"legend.loc"
[21] "legend.pars" "events"      "blocks"        "nc"
"nr"
```

We draw a simple graph of time series, as in Figure 2.10.

```
> data(sample_matrix)
> sample_xts <- as.xts(sample_matrix)
> plot(sample_xts[,1])
> class(sample_xts[,1])
[1] "xts" "zoo"
```

It can be seen from Figure 2.10 that xtsExtra::plot.xts() achieves a different effect from xts::plot.xts(). Next, we'll draw some more complex graphs.

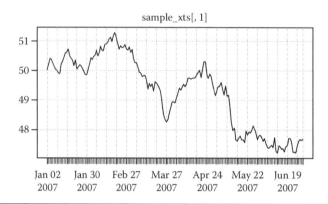

Figure 2.10   Time series.

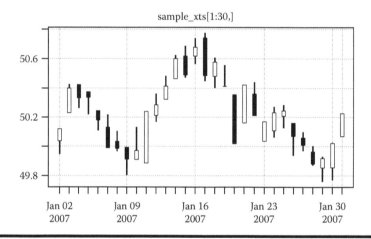

Figure 2.11   K-line chart drawn using xtsExtra.

2.3.3.1 K-Line Chart

We'll use gray and white as the default colors when we draw the K-line chart, as in Figure 2.11.

```
> plot(sample_xts[1:30,], type = "candles")
```

K-line chart in customized colors: Figure 2.12.

```
> plot(sample_xts[1:30,], type = "candles", bar.col.up = "blue",
bar.col.dn = "violet", candle.col = "green4")
```

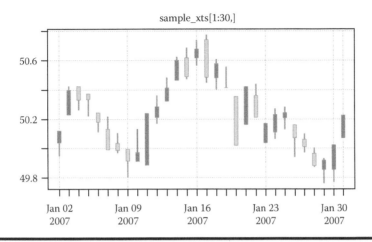

Figure 2.12   K-line chart in customized colors.

2.3.3.2 Configuration on Panel

Draw a basic panel, as in Figure 2.13.

```
> plot(sample_xts[,1:2])
```

Draw a multirow panel, as in Figure 2.14.

```
> plot(sample_xts[,rep(1:4, each = 3)])
```

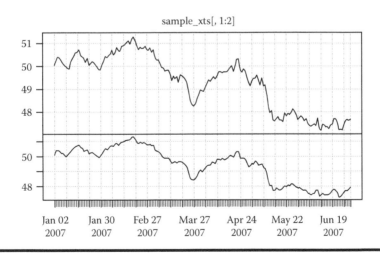

Figure 2.13   Basic panel.

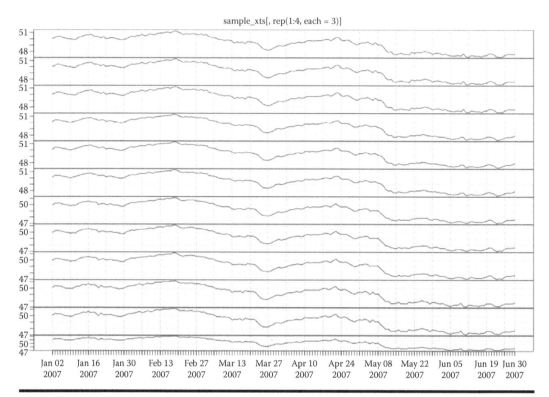

Figure 2.14   Multirow panel.

Draw a panel of free combination, as in Figure 2.15.

```
> plot(sample_xts[,1:4], layout.screens = matrix(c(1,1,1,1,2,3,4,4),
ncol = 2, byrow = TRUE))
```

Through drawing these graphs, we may find that plot.xts() provides various parameter configurations to help us achieve richer forms of expression in visualizing time series.

2.3.3.3 Configuration on Screens

Draw a graph display by double screen, with two lines in each screen as in Figure 2.16.

```
> plot(sample_xts, screens = 1:2)
```

Draw a graph displayed by a double screen and assign the screen and color of each line, as in Figure 2.17.

```
> plot(sample_xts, screens = c(1,2,1,2), col = c(1,3,2,2))
```

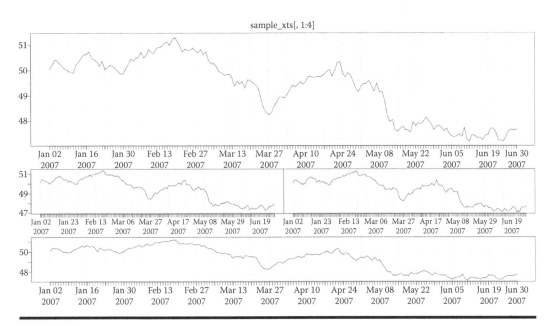

Figure 2.15 **Panel of free combination.**

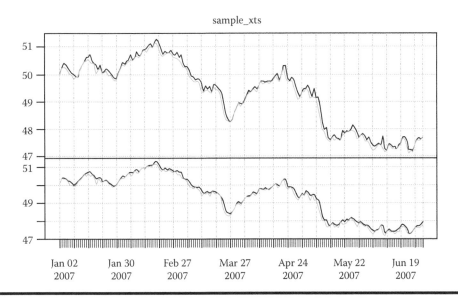

Figure 2.16 **Double screen display.**

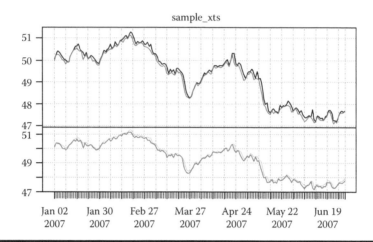

Figure 2.17   Double screen display with customized colors.

Draw a graph displayed by a double screen and assign different coordinates, as in Figure 2.18.

```
> plot(10^sample_xts, screens = 1:2, log=c("","y"))
```

Draw a graph displayed by a double screen and assign a different type of output, as in Figure 2.19.

```
> plot(sample_xts[,c(1:4, 3:4)], layout =
matrix(c(1,1,1,1,2,2,3,4,5,6), ncol = 2, byrow = TRUE), yax.loc =
"left")
```

Draw a graph displayed by multiple screens and set them in different groups as in Figure 2.20.

```
plot(sample_xts[,c(1:4, 3:4)], layout =
matrix(c(1,1,1,1,2,2,3,4,5,6), ncol = 2, byrow = TRUE), yax.loc =
"left")
```

2.3.3.4 Configuration on Events

Draw a dividing line of basic events, as in Figure 2.21.

```
> plot(sample_xts[,1], events = list(time = c("2007-03-15","2007-05-
01"), label = "bad days"), blocks = list(start.time = c("2007-03-
05", "2007-04-15"), end.time = c("2007-03-20","2007-05-30"), col =
c("lightblue1", "lightgreen")))
```

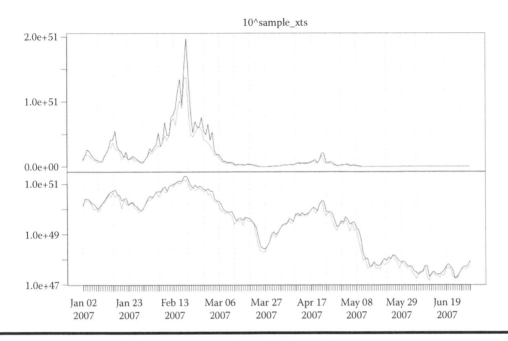

Figure 2.18 Double screen display with different coordinates.

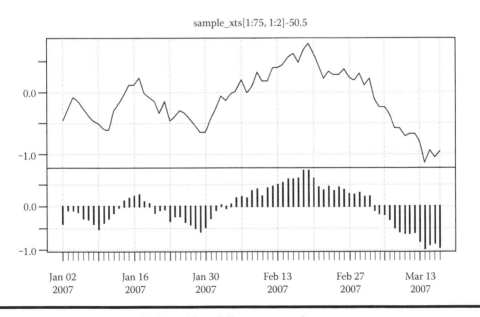

Figure 2.19 Double screen display with a different type of output.

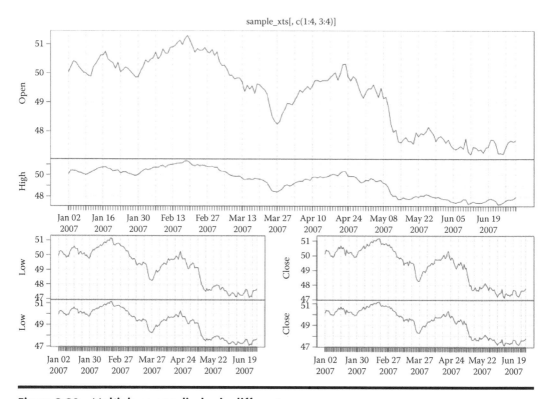

Figure 2.20   Multiple screen display in different groups.

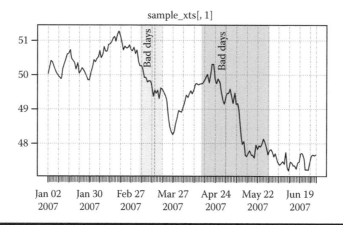

Figure 2.21   Dividing line of events.

2.3.3.5 Time Series of Double Coordinates

Draw a view of double coordinates, as in Figure 2.22.

```
> plot(sample_xts[,1],sample_xts[,2])
```

Draw a gradient view of double coordinates, as in Figure 2.23.

```
> cr <- colorRampPalette(c("#00FF00","#FF0000"))
> plot(sample_xts[,1],sample_xts[,2], xy.labels = FALSE, xy.lines =
TRUE, col = cr(NROW(sample_xts)), type = "l")
```

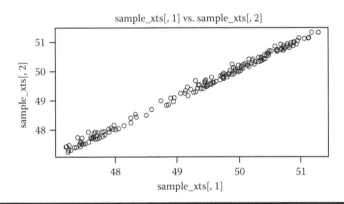

Figure 2.22   Double coordinates view.

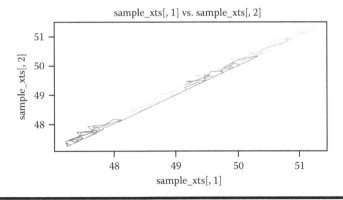

Figure 2.23   Gradient view of double coordinates.

2.3.3.6 Convert the xts Class and Draw Graphics

Draw a graph of data of the ts class, as in Figure 2.24.

```
> tser <- ts(cumsum(rnorm(50, 0.05, 0.15)), start = 2007, frequency
= 12)
> class(tser)
[1] "ts"
> plot(tser)
```

Draw a graph of the xts class and add background coordinate lines automatically, as in Figure 2.25.

```
> plot.xts(tser)
```

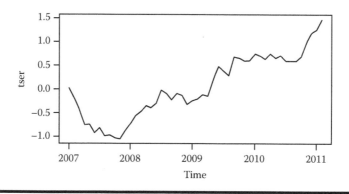

Figure 2.24   Graphic of data of the ts class.

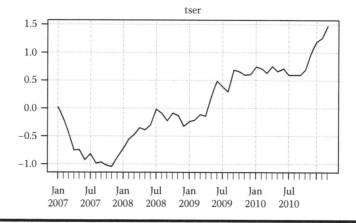

Figure 2.25   Graph of the xts class.

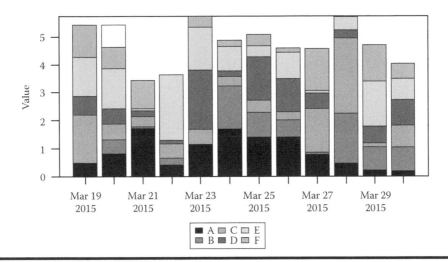

Figure 2.26   Bar chart.

2.3.3.7 Draw Graphs by Barplot

Use barplot() to draw a bar chart, as in Figure 2.26.

```
> x <- xts(matrix(abs(rnorm(72)), ncol = 6), Sys.Date() + 1:12)
> colnames(x) <- LETTERS[1:6]
> barplot(x)
```

We've seen the powerful drawing function of xtsExtra::plot.xts. It's easy to use to make graphics of time series with rich elements!

Chapter 3

Performance Monitoring Packages of R

This chapter mainly introduces three tool packages related to the performance of R, which may help readers find bottlenecks of performance in programs.

3.1 Local Cache Tool of R: memoise

Question

How do we strengthen the performance of R?

Local Cache Tool of R
http://blog.fens.me/r-cache-memoise

The caching technique is widely applied in computer systems. For applications of high concurrency access, it is the best solution to strengthen performance considering price effectiveness. Especially for repetitive calculation, cache can save a large amount of time for central processing unit (CPU), even up to 99%. Optimization is being conducted and changes have started. R leaders, represented by Hadley Wickham, are making R much faster.

3.1.1 Introduction to memoise

memoise is a simple cache package. Based on local cache, memoise reduces the repetitive calculation of a single computer and indirectly raises the CPU performance of a single computer. When we run a second calculation on a same function with the same parameters, it'll save CPU a great deal of time if we simply use the result of the first calculation as the result of the second calculation instead of running the calculation again.

The API of memoise is quite simple and contains only two functions.

- memoise: define memoised function, i.e., load the result of function calculation into local cache
- forget: forget past results; resets the cache of a memoised function

3.1.2 Installation of memoise

System environment used in this section:

- Windows 7 64bit
- R: 3.0.1 x86_64-w64-mingw32/x64 b4bit

Note: Installation of memoise supports both Windows 7 and Linux.
Installation of memoise is as follows:

```
# Start R.
~ R


# Install memoise.
> install.packages("memoise")
trying URL 'http://mirror.bjtu.edu.cn/cran/bin/windows/contrib/3.0/
memoise_0.1.zip'
Content type 'application/zip' length 10816 bytes (10 Kb)
opened URL
downloaded 10 Kb

package 'memoise' successfully unpacked and MD5 sums checked

# Load memoise.
> library(memoise
```

3.1.3 Use of memoise

Now we run a cache experiment. Suppose the running of a function costs 1 second of CPU.

```
# Define memoised function. Time stops for 1 second when the
function is ran.
> fun <- memoise(function(x) {Sys.sleep(1); runif(1)})
```

```
# First execution of fun(). The process costs 1 second.
> system.time(print(fun()))
[1] 0.05983416
User System Passing
   0    0    1

# Second execution of fun(). The process is being completed
immediately when result is loaded from cache.
> system.time(print(fun()))
[1] 0.05983416
User System Passing
   0    0    0

# Reset memoised function.
> forget(fun)
[1] TRUE

# Thrid execution of fun(). The process costs 1 second.
> system.time(print(fun()))
[1] 0.6001663
User System Passing
   0    0    1

# Fourth execution of fun(). The process is being completed
immediately when result is loaded from cache.
> system.time(print(fun()))
[1] 0.6001663
User System Passing
   0    0    0
```

The preceding execution process can be described as follows:

1. Create a memoised function fun() using memoise().
2. First execution of fun(). The process costs 1 second.
3. Second execution of fun(). The process is completed immediately when the result is loaded from the cache.
4. Clear the result of fun() in the cache by forget().
5. Third execution of fun(). Because the result of fun() has been cleared, so the process costs 1 second again.
6. Fourth execution of fun(). The process is completed immediately when the result is loaded from the cache.

3.1.4 Source Code Analysis of memoise()

The memoise project is a typical R project achieved by object-oriented programming. Object-oriented programming of R will be introduced in detail in the next book of this series, *R for Programmers: Advanced Techniques*. The source code of memoise can be found in Github at https://www.github.com/hadley/memoise.

3.1.4.1 Source Code of memoise()

Next we'll analyze the source code of memoise(). It mainly consists of three parts.

1. New_cache() creates new cache space for f().
2. Generate hash value of f() as key.
3. Return f() after caching.

```
# Define memoise and memoize.
memoise <- memoize <- function(f) {

  # Create new cache space.
  cache <- new_cache()

  # Generate hash value.
  memo_f <- function(...)  {
    hash <- digest(list(...))

    if (cache$has_key(hash))  {
      cache$get(hash)
    }  else  {
      res <- f(...)
      cache$set(hash, res)
      res
    }
  }
  attr(memo_f, "memoised") <- TRUE
  return(memo_f)
}
```

3.1.4.2 Source Code of forget()

The source code of forget() mainly consists of two parts.

1. Check if f() is cached in the environment.
2. If there exists a cache of f(), then clear the cache value of f().

```
forget <- function(f)  {

  # Check if function is cached.
  if (!is.function(f)) return(FALSE)

  env <- environment(f)
  if (!exists("cache", env, inherits = FALSE)) return(FALSE)

  # Clear the cache value of function.
  cache <- get("cache", env)
  cache$reset()

  TRUE
}
```

3.1.4.3 Private Function: Source Code of new_cache()

new_cache() mainly consists of three parts.

1. Define the cache object in new_cache() and save it in env.
2. Construct an object of class list by new_cache(), by measures including reset, set, get, has_ key, keys.
3. Access the cache object by the list object.

```
new_cache <- function()  {

  # Define cache object.
  cache <- NULL

  # Define cache reset function.
  cache_reset <- function() {
    cache <<- new.env(TRUE, emptyenv())
  }

  # Define cache set function.
  cache_set <- function(key, value)  {
    assign(key, value, env = cache)
  }

  # Define cache get function.
  cache_get <- function(key) {
    get(key, env = cache, inherits = FALSE)
  }

  # Check if cache exists.
  cache_has_key <- function(key) {
    exists(key, env = cache, inherits = FALSE)
  }

  # Clear cache.
  cache_reset()

  # reture the object of cache.
  list(
    reset = cache_reset,
    set = cache_set,
    get = cache_get,
    has_key = cache_has_key,
    keys = function() ls(cache)
  )
}
```

From the source code we not only can appreciate the design philosophy of memoise, but also gain a deep understanding of the writer of memoise on R. With a concise and effective code, memoise is worthwhile to learn.

3.2 Performance Monitoring Tool of R: Rprof

Question

How do we find bottlenecks of performance?

Performance Monitoring Tool
http://blog.fens.me/r-perform-rprof-profr/

As R becomes more and more widely used, its problem of calculation performance attracts increasing attention. How to get the cost of time of an algorithm on CPU clearly would be a key element to optimize performance. It's fortunate that the basic library of R has provided us with such a function for performance monitoring, Rprof().

3.2.1 Introduction to Rprof()

Rprof() is a log function of performance data of the core packages of R. It can print out the call relation of functions and the data of CPU cost of time. Then summaryRprof() will analyze the log data generated by Rprof() and produce a performance report. Finally, plot() in the profr library will visualize the report.

3.2.2 Definition of Rprof()

The system environment used in this section is

- Windows 7 64bit
- R: 3.0.1 x86_64-w64-mingw32/x64 b4bit

Note: Rprof and profr support both Windows 7 and Linux.

Rprof() is defined in the basic package utils. It can be directly used, without installation. Now view the definition of Rprof().

```
# Start R.
~ R

# View the definition of Rprof().
> Rprof
function (filename = "Rprof.out", append = FALSE, interval = 0.02,
memory.profiling = FALSE, gc.profiling = FALSE, line.profiling = FALSE,
numfiles = 100L, bufsize = 10000L)
{
```

```
if (is.null(filename))
filename <- ""
invisible(.External(C_Rprof, filename, append, interval,
memory.profiling, gc.profiling, line.profiling, numfiles,
bufsize))
}
<bytecode: 0x000000000d8efda8>
```

Rprof() is used to generate a log file to record performance indicators. Normally we just need to assign the filename.

3.2.3 Use of Rprof(): A Case Study of Stock Data Analysis

Consider stock data of Section 6.6 of this book as the test set. The data file is 000000_0.txt, 1.38MB. It can be found in the code files of this book. This section only serves as a performance test. The business implication of the data are explained further in Section 6.6.

```
> bidpx1<-read.csv(file="000000_0.txt",header=FALSE)
> names(bidpx1)<-c("tradedate","tradetime","securityid","bidpx1","bi
dsize1","offerpx1","offersize1")
> bidpx1$securityid<-as.factor(bidpx1$securityid)

> head(bidpx1)
  tradedate tradetime securityid bidpx1 bidsize1 offerpx1 offersize1
1  20130724    145004     131810  2.620     6960    2.630      13000
2  20130724    145101     131810  2.860    13880    2.890       6270
3  20130724    145128     131810  2.850   327400    2.851       1500
4  20130724    145143     131810  2.603    44630    2.800      10650
5  20130724    144831     131810  2.890    11400    3.000      77990
6  20130724    145222     131810  2.600  1071370    2.601      35750

> object.size(bidpx1)
1299920 bytes
```

Detailed explanation of fields: bidpx1 is the price of the best bid. Bidsize1 is the size of the best bid. Offerpx1 is the price of the best offer. Offersize1 is the size of the best offer.

Task of calculation: divide the data into groups by securityid and calculate the mean of the price of the best bid and the total amount of the size of the best bid.

```
# Load plyr package for data processing.
> library(plyr)

# Encapsulate the data processing to fun1().
> fun1<-function(){
+     datehour<-paste(bidpx1$tradedate,substr(bidpx1$tradetime,1,2),
sep="")
+     df<-cbind(datehour,bidpx1[,3:5])
```

```
+         ddply(bidpx1,.(securityid,datehour),summarize,price=mean(bi
dpx1),size=sum(bidsize1))
+ }

> head(fun1())
  securityid   datehour     price       size
1     131810 2013072210 3.445549 189670150
2     131810 2013072211 3.437179 131948670
3     131810 2013072212 3.421000       920
4     131810 2013072213 3.509442 299554430
5     131810 2013072214 3.578667 195130420
6     131810 2013072215 1.833000    718940
```

Check the running time of fun1 by system.time(). Run the operation twice, we may find that the time cost of the system is similar and that there is no cache for the second operation.

```
> system.time(fun1())
User System Passing
0.08 0.00 0.07

> system.time(fun1())
User System Passing
0.06 0.00 0.06
```

Use Rprof() to record data of performance indicators and output it to file.

```
# Define the position of output of performance log.
> file<-"fun1_rprof.out"

# Start performance monitoring.
> Rprof(file)

# Execute the function of calculation.
> fun1()

# Stop performance monitoring and output the result to file.
> Rprof(NULL)
```

Check the generated file of performance indicators: fun1_rprof.out.

```
~ vi fun1_rprof.out

sample.interval=20000
"substr" "paste" "fun1"
"paste" "fun1"
"structure" "splitter_d" "ddply" "fun1"
```

```
".fun" "" ".Call" "loop_apply" "llply" "ldply" "ddply" "fun1"
".fun" "" ".Call" "loop_apply" "llply" "ldply" "ddply" "fun1"
".fun" "" ".Call" "loop_apply" "llply" "ldply" "ddply" "fun1"
"[[" "rbind.fill" "list_to_dataframe" "ldply" "ddply" "fun1"
```

In fact, we cannot understand this log. So we need to use summaryRprof() to explain this log. Check the statistical report by summaryRprof().

```
# Load the file.
> summaryRprof(file)
$by.self
                self.time self.pct total.time total.pct
".fun"              0.06    42.86       0.06     42.86
"paste"             0.02    14.29       0.04     28.57
"[["                0.02    14.29       0.02     14.29
"structure"         0.02    14.29       0.02     14.29
"substr"            0.02    14.29       0.02     14.29

$by.total
                     total.time total.pct self.time self.pct
"fun1"                     0.14    100.00      0.00     0.00
"ddply"                    0.10     71.43      0.00     0.00
"ldply"                    0.08     57.14      0.00     0.00
".fun"                     0.06     42.86      0.06    42.86
".Call"                    0.06     42.86      0.00     0.00
""                         0.06     42.86      0.00     0.00
"llply"                    0.06     42.86      0.00     0.00
"loop_apply"               0.06     42.86      0.00     0.00
"paste"                    0.04     28.57      0.02    14.29
"[["                       0.02     14.29      0.02    14.29
"structure"                0.02     14.29      0.02    14.29
"substr"                   0.02     14.29      0.02    14.29
"list_to_dataframe"        0.02     14.29      0.00     0.00
"rbind.fill"               0.02     14.29      0.00     0.00
"splitter_d"               0.02     14.29      0.00     0.00

$sample.interval
[1] 0.02

$sampling.time
[1] 0.14
```

Explanation of data:

■ $by.self: cost of time of current function. Self.time is the time of actual running, and total.time is the accumulated time of running.
■ $by.total: overall situation of function call. Self.time is the time of actual running, and total.time is the accumulated time of running.

We can find from $by.self that most of the time is spent on .fun.

- .fun: the time of actual running is 0.06, accounting for 42.86% of the time of current function
- Paste: the time of actual running is 0.02, accounting for 14.29% of the time of current function
- "[[": the time of actual running is 0.02, accounting for 13.29% of the time of current function
- "structure": the time of actual running is 0.02, accounting for 14.29% of the time of current function
- "substr": the time of actual running is 0.02, accounting for 14.29% of the time of current function

In $by.total, the result is ordered sequence of execution

- 4 fun1: the accumulated time of running is 0.14, accounting for 100% of total accumulated time of running. The time of actual running is 0.00.
- 3.fun: the accumulated time of running is 0.06, accounting for 42.86% of total accumulated time of running. The time of actual running is 0.06.
- 2 paste: the accumulated time of running is 0.04, accounting for 28.57% of total accumulated time of running. The time of actual running is 0.02.
- 1 splitter_d: the accumulated time of running is 0.02, accounting for 14.297% of total accumulated time of running. The time of actual running is 0.00.

Now we know the CPU time of every function called. To optimize performance, we'll start from the function that cost the most time.

3.2.4 Use of Rprof(): A Case Study of Data Download

First, we need to install the stockPortfolio package and download stock data through stockPortfolio. Then we use Rprof() to monitor the cost of time of the downloading.

```
# Install stockPortfolio.
> install.packages("stockPortfolio")

# Load stockPortfolio.
> library(stockPortfolio)
> fileName <- "Rprof2.log"

# Start performance monitoring.
> Rprof(fileName)

# Download the stock data of Google, Microsoft and IBM.
> gr <- getReturns(c("GOOG", "MSFT", "IBM"), freq="week")
> gr
Time Scale: week
Average Return
        GOOG         MSFT          IBM
0.004871890 0.001270758 0.001851121

# Finish performance monitoring.
> Rprof(NULL)
```

```
# Check performance report.
> summaryRprof(fileName)$by.total[1:8,]
                        total.time total.pct self.time self.pct
"getReturns"                 6.76    100.00      0.00     0.00
"read.delim"                 6.66     98.52      0.00     0.00
"read.table"                 6.66     98.52      0.00     0.00
"scan"                       4.64     68.64      4.64    68.64
"file"                       2.02     29.88      2.02    29.88
"as.Date"                    0.08      1.18      0.02     0.30
"strptime"                   0.06      0.89      0.06     0.89
"as.Date.character"          0.06      0.89      0.00     0.00
```

From the eight printed records that cost the most time, we can see that most of the time of actual running (self.time) is spent on file:2.02, scan:4.64.

3.2.5 To Visualize Performance Indicators Using Profr Package

Profr package provides a function of visualized display on the performance report output by Rprof(). It is more user-friendly and easier to read than the pure-text report of summaryRprof(). First, we need to install profr.

```
# Install profr.
> install.packages("profr")

# Load profr.
> library(profr)

# Use ggplot2 to draw graphics.
# Load ggplot2.
> library(ggplot2)
```

The first case of data visualization is stock data analysis. The following code will generate Figures 3.1 and 3.2.

```
> file<-"fun1_rprof.out"

# Graphic by plot.
> plot(parse_rprof(file))

# Graphic by ggplot2.
> ggplot(parse_rprof(file))
```

The second case of data visualization is data download. The following code will generate Figures 3.3 and 3.4.

```
> fileName <- "Rprof2.log"
> plot(parse_rprof(fileName))
> ggplot(parse_rprof(fileName))
```

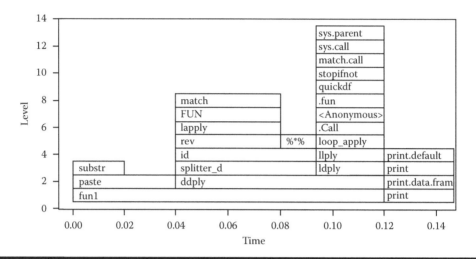

Figure 3.1   Graphic generated by the plot.

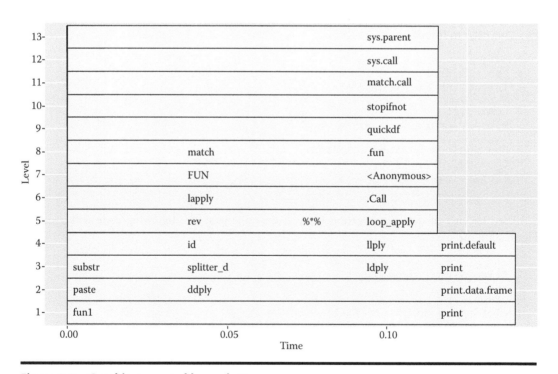

Figure 3.2   Graphic generated by ggplot2.

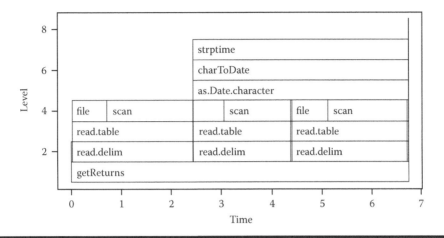

Figure 3.3   Graphic drawn by the plot.

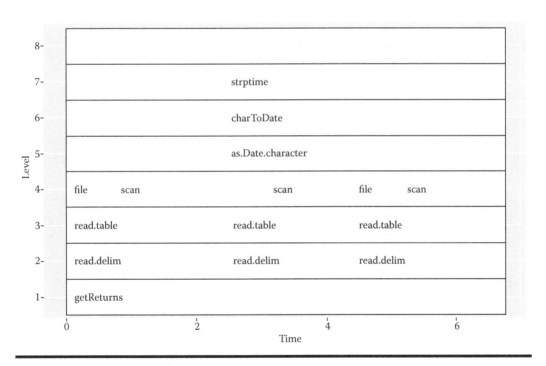

Figure 3.4   Graphic drawn by ggplot2.

3.2.6 Use of Command Line of Rprof

The command line of Rprof can be used to check log files conveniently.

3.2.6.1 Check the Help File of Command Line of Rprof

```
~ D:\workspace\R\preforemence\Rprof>R CMD Rprof —help
Usage: R CMD Rprof [options] [file]

Post-process profiling information in file generated by Rprof().

Options:
  -h, —help        print short help message and exit
  -v, —version     print version info and exit
  —lines           print line information
  —total           print only by total
  —self            print only by self
  —linesonly       print only by line (implies —lines)
  —min%total=      minimum% to print for 'by total'
  —min%self=       minimum% to print for 'by self'

If 'file' is omitted 'Rprof.out' is used

Report bugs at bugs.r-project.org.
```

3.2.6.2 Use of Command Line of Rprof

Here is the completed report:

```
~ D:\workspace\R\preforemence\Rprof>R CMD Rprof fun1_rprof.out

Each sample represents 0.02 seconds.
Total run time: 0.14 seconds.

Total seconds: time spent in function and callees.
Self seconds: time spent in function alone.

    %      total      %      self
 total    seconds   self   seconds   name
 100.0     0.14      0.0     0.00     "fun1"
  71.4     0.10      0.0     0.00     "ddply"
  57.1     0.08      0.0     0.00     "ldply"
  42.9     0.06     42.9     0.06     ".fun"
  42.9     0.06      0.0     0.00     ".Call"
  42.9     0.06      0.0     0.00     ""
  42.9     0.06      0.0     0.00     "llply"
  42.9     0.06      0.0     0.00     "loop_apply"
  28.6     0.04     14.3     0.02     "paste"
  14.3     0.02     14.3     0.02     "[["
  14.3     0.02     14.3     0.02     "structure"
  14.3     0.02     14.3     0.02     "substr"
  14.3     0.02      0.0     0.00     "list_to_dataframe"
```

```
14.3      0.02      0.0      0.00      "rbind.fill"
14.3      0.02      0.0      0.00      "splitter_d"

  %       self       %      total
self     seconds   total   seconds    name
42.9      0.06      42.9     0.06      ".fun"
14.3      0.02      28.6     0.04      "paste"
14.3      0.02      14.3     0.02      "[["
14.3      0.02      14.3     0.02      "structure"
14.3      0.02      14.3     0.02      "substr"
```

Here is the report that displays only the parts in which indicator *total* accounts for more than 50% of the time.

```
~ D:\workspace\R\preforemence\Rprof>R CMD Rprof —total —min%total=50
fun1_rprof.out

Each sample represents 0.02 seconds.
Total run time: 0.14 seconds.

Total seconds: time spent in function and callees.
Self seconds: time spent in function alone.

  %       total      %      self
total    seconds   self    seconds    name
100.0     0.14      0.0      0.00      "fun1"
 71.4     0.10      0.0      0.00      "ddply"
 57.1     0.08      0.0      0.00      "ldply"
```

We can optimize the performance of code by using Rprof, and the computing power will no longer be the bottleneck.

3.3   Performance Visualization Tool of R: lineprof

Question
Is there a visualization performance monitoring tool?

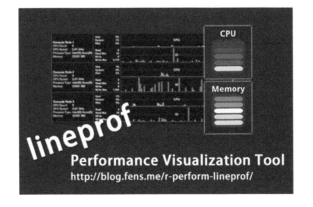

More and more people are starting to explore the field of data visualization. Images can be more expressive than words, and interactive images based on HTML offer better visualization than static PNG images. R has been fully prepared for data visualization. There are packages of image visualization, ggplot2; packages of world map visualization, ggmap; packages of stock visualization, quantmod; and packages of interactive visualization based on HTML, googleVis in R. We can turn data into images and even make the image dynamic by just entering a few lines of code. The performance report will be used as an entry point to introduce a visualization package of R, lineprof.

3.3.1 Introduction to lineprof

lineprof is a project of data visualization, aiming at visualizing the effect of performance monitoring in a more user-friendly way. Section 3.2 contains an output of the performance data in the form of graphs by using profr, but the graphs are all static. lineprof can do better than that: it will generate interactive webpage based on shiny and let users find problems by themselves.

lineprof is also a work of Hadley Wick. Now the project is published only on Github, at https://www.github.com/hadley/lineprof. There are two main categories of functions in the API of lineprof.

1. Performance function:
    a. focus: set the zoom of display height
    b. auto_focus: set the zoom of display height automatically
    c. lineprof: record the occupation of CPU and RAM
    d. shine: output by shiny
2. Inner function: ancillary function
    a. align: align the source code
    b. find_ex: load demo
    c. line_profile: output of formatted data of performance monitoring (Rprof)
    d. parse_prof: formatted output
    e. reduce_depth: set the depth of output

3.3.2 Installation of lineprof

System environment used in this section:

- Linux: Ubuntu Server 12.04.2 LTS 64bit
- R: 3.0.1 x86_64-pc-linux-gnu
- IP: 192.168.1.201

Note: lineprof only supports Linux.
Because lineprof hasn't been published on CRAN, we can only install lineprof from Github. We need to use devtools package to install projects of R from Github.

```
# Start R.
~ R

# Load devtools.
> library(devtools)
```

```
# Install lineprof through devtools.
> install_github("lineprof")

# Install customized console of shiny.
> install_github("shiny-slickgrid", "wch")

# Load lineprof.
> library(lineprof)
```

3.3.3 Use of lineprof

We use the official case to introduce the use of lineprof.

```
# Load script file, read-delim.r.
> source(find_ex("read-delim.r"))

# Load test set.
> wine <- find_ex("wine.csv")

# Execute read_delim algorithm, and record performance indicators
through lineprof.
> x <- lineprof(read_delim(wine, sep = ","), torture = TRUE)
Zooming to read-delim.r (97% of total time)
```

In the resource of this case, read-delim.r is the script file of objective function, wine.csv is the test set, and x:lineprof is the data report generated.

Check read-delim.r.

```
  function(file, header = TRUE, sep = ",", stringsAsFactors = TRUE)
{
  # Determine number of fields by reading first line
  first <- scan(file, what = character(1), nlines = 1, sep = sep,
quiet = TRUE)
  p <- length(first)

  # Load all fields
  all <- scan(file, what = as.list(rep("character", p)), sep = sep,
    skip = if (header) 1 else 0, quiet = TRUE)

  # Convert from strings to appropriate types
  all[] <- lapply(all, type.convert, as.is = !stringsAsFactors)

  # Set column names
  if (header)  {
    names(all) <- first
    rm(first)
  } else  {
    names(all) <- paste0("V", seq_along(all))
  }
```

```
  # Convert list into data frame
  class(all) <- "data.frame"
  attr(all, "row.names") <- c(NA_integer_, -length(all[[1]]))

  all
}
```

Load wine.csv.

```
> df<-read.csv(file=wine)

# Size of the dataset.
> object.size(df)
20440 bytes

# Display the first 3 rows of data.
> head(df, 3)
  type alcohol  malic   ash    alcalinity magnesium phenols flavanoids
  nonflavanoids proanthocyanins color  hue dilution  proline
1  A      14.23  1.71  2.43    15.6       127       2.80    3.06
     0.28 2.29   5.64   1.04    3.92       1065
2  A      13.20  1.78  2.14    11.2       100       2.65    2.76
     0.26 1.28   4.38   1.05    3.40       1050
3  A      13.16  2.36  2.67    18.6       101       2.80    3.24
     0.30 2.81   5.68   1.03    3.17       1185
```

X object: data report generated by lineprof.

```
# Object to record performance indicators.
> x
Reducing depth to 2 (from 8)
Common path:
      time alloc release dups  ref  src
1   0.002 0.001   0.000     0  #3   read_delim
2   0.049 0.009   0.003    11  #3   read_delim/scan
3   0.026 0.001   0.008     0  #4   read_delim
4   0.379 0.072   0.006    14  #7   read_delim/scan
5   0.003 0.000   0.000     0  #11  read_delim
6   0.106 0.015   0.030     3  #11  read_delim/lapply
7   0.008 0.004   0.000     3  #11  read_delim
8   0.210 0.028   0.077    36  #16  read_delim/rm
9   0.004 0.001   0.000     1  #22  read_delim
10  0.035 0.005   0.004     8  #23  read_delim/[[
11  0.002 0.000   0.000     1  #23  read_delim/length
12  0.001 0.000   0.000     1  #23  read_delim/c
13  0.006 0.004   0.000     1  #23  read_delim
14  0.001 0.000   0.000     0  #23  read_delim/attr<-
```

Use shinySlickgrid to visualize data of performance indicators and output the result in the form of a webpage.

```
# Load shinySlickgrid.
> library(shinySlickgrid)

# Start shiny.
> shine(x)
Loading required package: shiny
Shiny URLs starting with/lineprof will mapped to/home/conan/R/
x86_64-pc-linux-gnu-library/3.0/lineprof/www
Shiny URLs starting with/slickgrid will mapped to/home/conan/R/
x86_64-pc-linux-gnu-library/3.0/shinySlickgrid/slickgrid

Listening on port 6742
```

Shiny will open a Web server in the background. The default access port of Web is 6742. Remote access is available through browser. Open browser and type in http://192.168.1.201:6742 as in Figure 3.5.

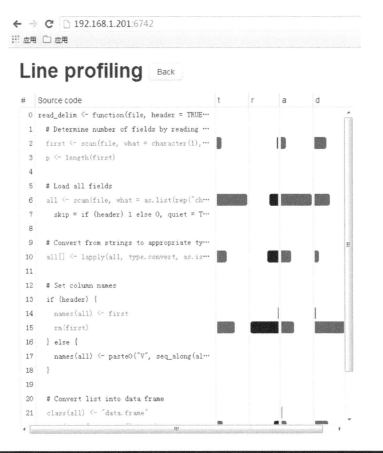

Figure 3.5 Visualization performance monitoring.

There are six rows in the table of webpage in Figure 3.5. # is the number of row, source code is the source code of monitored objective function, t is the total time (second) of the current executed row, r is the memory released, a is the memory allocated, and d is the times of duplicates. The following is the explanation of function performance data.

- #6: used for loading data. Total time: 0.309s. Memory allocated: 0.064mb. Times of duplicates: 14
- #15: used for clearing data. Total time: 0.179s. Memory allocated: 0.065mb. Times of duplicates: 37

By using the visualization tool, lineporf, we can produce more flexible, more intuitive, and better-looking reports, or even interactive reports of Web version. If you compare the effect with that of Section 3.2, you will likely be surprised by how powerful R is and how rapidly R can progress. R has nearly infinite potential, but it still needs more people to push forward its progress. I hope that you can also make contributions to the progress of R.

R SERVER

Chapter 4

Cross-Platform
Communication of R

This chapter mainly introduces four tool packages of cross-platform communication of R. It will help readers implement the communication between R, Java, and JavaScript.

4.1 Cross-Platform Communication between Rserve and Java

Question

How to call R in Java?

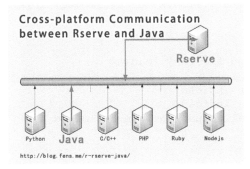

The current mainstream heterogeneous cross-platform communication component, Apache Thrift, has been widely adopted all over the world. It supports 15 programming languages, but R is not within the range. If we need to implement cross-platform communication of R, we need to find solutions from the official R community CRAN. Solutions of combining two languages, including rJava, Rcpp, and rpy, all choose to load an engine of R into the memory environment of other languages. The strength of such solutions is their high efficiency, while their weaknesses include tight coupling, limited extension, and nonreusable interface programs.

Rserve provides us with a new choice. It is an abstract network interface of R. It implements the communication of languages based on Transmission Control Protocol/Internet Protocol (TCP/IP) agreement. Through the program call of the Client/Server (C/S) structure, Rserve supports the communication between R and other languages including C/C++, Java, PHP, Python, Ruby, and Node.js. Rserve supports many functions including remote connection, user authentication, and file transfer. We can use R as the background service engine to process tasks including statistical modeling, data analysis, and plotting. We explore the cross-platform communication between Rserve and Java in this section.

4.1.1 Installation of Rserve

System environment used in this section:

- Linux Ubuntu 12.04.2 LTS 64bit server
- R: 3.0.1 x86_64-pc-linux-gnu
- IP: 192.168.1.201

Note: Rserve supports both Windows 7 and Linux. Because Rserve is mainly used as a communication server, Linux is recommended.

The installation of Rserve is as follows:

```
# Start R.
~ R

# Install Rserve.
> install.packages("Rserve")
installing via 'install.libs.R' to/usr/local/lib/R/site-library/
Rserve
** R
** inst
** preparing package for lazy loading
** help
*** installing help indices
** building package indices
** testing if installed package can be loaded
* DONE (Rserve)

# Start Rserve.
~ R CMD Rserve
R version 3.0.1 (2013-05-16) — "Good Sport"
Copyright (C) 2013 The R Foundation for Statistical Computing
Platform: x86_64-pc-linux-gnu (64-bit)

R is free software and comes with ABSOLUTELY NO WARRANTY.
You are welcome to redistribute it under certain conditions.
Type 'license()' or 'licence()' for distribution details.

  Natural language support but running in an English locale
```

```
R is a collaborative project with many contributors.
Type 'contributors()' for more information and
'citation()' on how to cite R or R packages in publications.

Type 'demo()' for some demos, 'help()' for on-line help, or
'help.start()' for an HTML browser interface to help.
Type 'q()' to quit R.

Rserv started in daemon mode.

# Check the process.
~ ps -aux|grep Rserve
conan    7142  0.0  1.2 116296 25240 ?           Ss    09:13    0:00/usr
/lib/R/bin/Rserve

# Check the port.
~ netstat -nltp|grep Rserve
tcp        0        0 127.0.0.1:6311              0.0.0.0:*
LISTEN     7142/Rserve
```

Here the server of Rserve has been started, with port 6311. 127.0.0.1 indicates that only local application access is available. If we want to access Rserve remotely, we need to open remote mode by adding parameter –RS-enable-remote to the startup command.

```
# Kill the daemon process of Rserve.
~ kill -9 7142

# Start Rserve by remote mode.
~ R CMD Rserve —RS-enable-remote

# Check the port.
~ netstat -nltp|grep Rserve
tcp        0        0 0.0.0.0:6311              0.0.0.0:*              LISTEN
7173/Rserve
```

0.0.0.0 indicates that restriction on IP access is lifted and we can access Rserve remotely from now on.

4.1.2 Remote Connection between Rserve and Java

Java is run directly in Windows 7 through EclipseIDE, which helps Java connect the Rserve server of Linux remotely. The Windows 7 environment is as follows:

- JAVA: Oracle SUN JDK 1.6.0_45 64bit
- Eclipse: Juno Service Release 2
- IP: 192.168.1.13

The IP of Linux server of Rserve is 192.168.1.201.

4.1.2.1 Download JAR Package of Java Client

We can download the JAR package of Java client, REngine.jar and RserveEngine.jar, through http://www.rforge.net/Rserve/files/. By using these two Jar packages, we can implement the communication between Java and Rserve.

- REngine.jar: used for mapping between data of class R and data of class Java
- RserveEngine.jar: used in communication program of Rserve

We can check the help files of these two libraries in the official file javadoc at http://rforge.net /org/doc/. These two JAR packages are binary files compiled by Java, and there is no source code file available.

4.1.2.2 Create Java Project in Eclipse

Create a new Java project in Eclipse and load Jar package, as in Figure 4.1.

4.1.2.3 Implementation of Java Programming

Then we write a Java file, Demo1.java. There are two ways to call Rserve server remotely using Java in Demo1.java.

- Main(): This is an entrance to start Java, by instantiating a demo object and calling callRserve().
- callRserve(): Create a socket link to access Rserve remotely and send two sentences of R to Rserve server in the form of character string. Then run calculation on Rserve, return the result and output it in Java.

The code of Demo1.java is as follows.

```java
package org.conan.r.rserve;
import org.rosuda.REngine.REXP;
import org.rosuda.REngine.REXPMismatchException;
import org.rosuda.REngine.Rserve.RConnection;
import org.rosuda.REngine.Rserve.RserveException;


public class Demo1 {

    /**
     * Start Java by Main.
     */
    public static void main(String[] args) throws RserveException,
REXPMismatchException {
        Demo1 demo = new Demo1();
        demo.callRserve();
    }
```

```
    /**
     * Access Rserve.
     */
    public void callRserve() throws RserveException,
REXPMismatchException  {
        //Create access connect.
        RConnection c = new RConnection("192.168.1.201");
        //Run a R sentence.
        REXP x = c.eval("R.version.string");
        //Print outthe result in Java.
        System.out.println(x.asString());
        //Run rnorm(10).
        double[] arr = c.eval("rnorm(10)").asDoubles();
        //Print out the result in loop.
        for (double a: arr) {
            System.out.print(a + ",");
        }
    }
}
```

The result of running is as follows:

```
R version 3.0.1 (2013-05-16)
1.7695224124757984,-0.29753038160770323,0.26596993631142246,
1.4027325257239547,-0.30663565983302676,-0.17594309812158912,
0.10071253841443684,0.9365455161259986,0.1127211943643970l,
0.5766373030674361,
```

Thus we've implemented the communication between Java and R through Rserve easily. To speak more precisely, we've implemented a communication based on TCP/IP by using Java to access the Rserve server. After we solve the problem of communication, we may make full use of our imagination to apply R in more fields. This section is only a simple introduction on the installation and launch of Rserve. For the detailed use and configuration of Rserve server, please check Section 6.1.

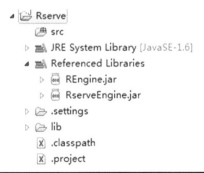

Figure 4.1　Java project in Eclipse.

4.2 Rsession Makes It Easier for Java to Call R

Question

Is there an easier way for Java to call R?

Rserve, as an important communication interface of R, has become an important channel for the extension of R. But as Rserve is based on underlying structure, it is rather difficult to call the API interface of Rserve using Java. Such a circumstance leads to the creation of Rsession. Rsession, being an encapsulation of Rserve, provides higher API interfaces, including Rserve server control and multisession mechanism, and supports Windows. Rsession makes it easier for Java to call API of Rserve and provides a simpler way for Java to access remote or local Rserve instances, whereas the other library for communication between R and Java, JRI, does not support a multisession mechanism. We introduce JRI in the next section.

4.2.1 Download of Rsession

System environment used in this section:

- Windows 7 64bit
- R: 3.0.1 x86_64-w64-mingw32/x64 b4bit

Note: Rsession supports both Windows 7 and Linux.

4.2.1.1 Download the Distribution Directly

Rsession includes three jar packages: REngine.jar, Rserve.jar and Rsession.jar. We can download the distribution through http://rsession.googlecode.com/svn/trunk/Rsession/. It can be used directly after uncompress.

4.2.1.2 Download the Source Code and Compile Distribution

We can download Rsession using source code. The SVN address of Rsession is http://rsession.googlecode.com/svn/trunk/Rsession/. The following is the operation to download SVN from Rsession.

```
# Enter local directory.
~ cd d:\workspace\java

# Download through SVN.
~ svn checkout http://rsession.googlecode.com/svn/trunk/
rsession-read-only

# Change the name of directory.
~ mv rsession-read-only rsession

# Run directory of Rsession code.
~ cd rsession\Rsession
```

Rsession, constructed by Ant, can be compiled and packaged by us. Ant is an automatic constructing tool of Java.

```
# Use ant to construct project.
~ ant
Buildfile: d:\workspace\java\rsession\Rsession\build.xml

clean:

clean-dist:

init:
    [mkdir] Created dir: d:\workspace\java\rsession\Rsession\build
    [mkdir] Created dir: d:\workspace\java\rsession\Rsession\dist\lib

resource:
     [copy] Copying 28 files to d:\workspace\java\rsession\Rsession\
dist\lib
     [copy] Copied 12 empty directories to 1 empty directory under
d:\workspace\java\rsession\Rsession\dist\lib

compile:
    [javac] d:\workspace\java\rsession\Rsession\build.xml:33:
warning: 'includeantruntime' was not set, defaulting to bu
ild.sysclasspath=last; set to false for repeatable builds
    [javac] Compiling 10 source files to d:\workspace\java\rsession\
Rsession\build

dist:
      [jar] Building jar: d:\workspace\java\rsession\Rsession\dist\
lib\Rsession.jar
      [zip] Building zip: d:\workspace\java\rsession\Rsession\dist\
libRsession.zip

BUILD SUCCESSFUL
Total time: 2 seconds
```

By running the ant commands given in the preceding, we can generate a distribution, libRsession.zip, in d:\workspace\java\rsession\Rsession\dist\. There is a jmatharray.jar in the manually compiled distribution, which does not exist in the direct downloaded distribution. I think this jar file might be a dependent package needed in the compiling process. It's not needed in the running, and will not impact the running process.

4.2.2 Construct Rsession Projects Using Eclipse

Copy the directory document Rsession\dist\ to the project, and load the project to environment variable, as in Figure 4.2.

4.2.3 API Introduction of Rsession

Check the class library Rsession.jar as in Figure 4.3. All the API of Rsession, including interface, function, and auxiliary, can be found in it. Here is an introduction to each kind.

1. Interface
   - BusyListener: to listen the condition of R engine
   - EvalListener: to listen the running of R script of R engine
   - Logger: for journal output
   - UpdateObjectsListener: to listen the changes of environment when R is run

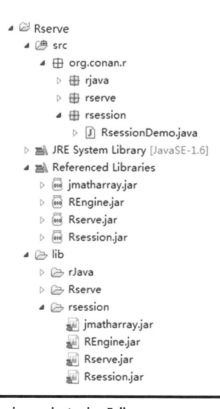

Figure 4.2   Construct a Rsession project using Eclipse.

Figure 4.3 File structure of Rsession.jar.

2. Function
   - Rdaemon: daemon process of RServe
   - RLogPanel: display the space of R journal
   - RObjectsPanel: a control to display R variables
   - RserverConf: connect configuration documents of Rserve instances
   - Rsession: connect Rserve instances
   - StartRserve: start local Rserve

4.2.4 Use of Rsession

Then we use Rsession to remotely access the server of Rserve.

4.2.4.1 Server Environment of Rserve

System environment for remote server:

- Linux: Ubuntu 12.04.2 LTS 64bit
- R: 3.0.1 x86_64-pc-linux-gnu

Remote Rserve environment:

- Rserve: Rserve v1.7-1
- IP: 192.168.1.201, allows remote access
- Interface: 6311
- Login authentication: username:conan, password:conan
- Character encoding: utf-8

The environment configuration of the server of Rserve is the same as that of RScilent of Rserve in Section 5.2.

4.2.4.2 Java Code

Then we'll write a class of Java to call the remote Rserve server.

Java code: RsessionDemo.java

```java
package org.conan.r.rsession;

import java.io.File;
import java.util.Properties;

import org.math.R.RserverConf;
import org.math.R.Rsession;
import org.rosuda.REngine.REXPMismatchException;

public class RsessionDemo {

    /**
     * "Main" to start java.
     */
    public static void main(String args[]) throws
REXPMismatchException {

        //Establish remote connection.
        RserverConf rconf = new RserverConf("192.168.1.201", 6311,
"conan", "conan", new Properties());
        Rsession s = Rsession.newInstanceTry(System.out, rconf);

        //Run R script.
        double[] rand = s.eval("rnorm(5)").asDoubles();
        System.out.println(rand);
        // Create an R object.
        s.set("demo", Math.random());
        s.eval("ls()");

        // Save the running condition of R to file.
        s.save(new File("./output/save.Rdata"), "demo");

        // Delete demo.
        s.rm("demo");
        s.eval("ls()");

        // Load R environment from file.
        s.load(new File("./output/save.Rdata"));
        s.eval("ls()");
        s.eval("print(demo)");

        //Create a data.frame object.
        s.set("df", new double[][] {{1, 2, 3}, {4, 5, 6}, {7, 8, 9},
{10, 11, 12}}, "x1", "x2", "x3");
        double df$x1_3 = s.eval("df$x1[3]").asDouble();
        System.out.println(df$x1_3);
        s.rm("df");
```

```
            //Generate an image file.
            s.eval("getwd()");
            s.toJPEG(new File("./output/plot.png"), 400, 400,
    "plot(rnorm(10))");

            //Output in the form of HTML.
            String html = s.asHTML("summary(rnorm(100))");
            System.out.println(html);

            // Output in the form of text.
            String txt = s.asString("summary(rnorm(100))");
            System.out.println(txt);

            // Install new class library.
            System.out.println(s.installPackage("sensitivity", true));

            s.end();
        }
    }
```

4.2.4.3 Run Journal Output

I'll display the related Java program with the journal together for the convenience of reading. First run the R script.

```
// Run R script.
double[] rand = s.eval("rnorm(5)").asDoubles();
for(double ran:rand){
    System.out.print(ran+",");
}

// Journal output.
[eval] rnorm(5)

  org.rosuda.REngine.REXPDouble@5f934ad[5]
{0.08779203903807914,0.039929482749452114,-0.8788534039223883,
-0.8875740206608903,-0.8493446334021442}
0.08779203903807914,0.039929482749452114,-0.8788534039223883,
-0.8875740206608903,-0.8493446334021442
```

Then create an R object and save the R environment.

```
// Create an R object.
s.set("demo", Math.random());
s.eval("ls()");

// Save the R environment to local file.
s.save(new File("./output/save.Rdata"), "demo");
```

```
// Delete demo.
s.rm("demo");
s.eval("ls()");

// Load R environment from file.
s.load(new File("./output/save.Rdata"));
s.eval("ls()");
s.eval("print(demo)");

// Journal output.
[set] demo
```

Create a data.frame object.

```
s.set("df", new double[][] {{1, 2, 3}, {4, 5, 6}, {7, 8, 9}, {10, 11,
12}}, "x1", "x2", "x3");
double df$x1_3 = s.eval("df$x1[3]").asDouble();
System.out.println(df$x1_3);
s.rm("df");
// Journal output
[set] df
```

Generate a local image file, as in Figure 4.4.

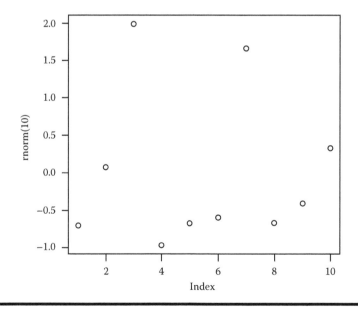

Figure 4.4   toJEPG image.

```
s.toJPEG(new File("./output/plot.png"), 400, 400, "plot(rnorm(10))");
// Journal Output
[set] plotfile_1100539400
```

Output in the form of HTML.

```
String html = s.asHTML("summary(rnorm(100))");
System.out.println(html);

// Journal output.
<html>    Min.    1st Qu.    Median    Mean    3rd Qu.    Max. <br/>
-2.332000  -0.659900  0.036920  0.004485  0.665800    2.517000 </html>
```

Output in the form of text.

```
String txt = s.asString("summary(rnorm(100))");//format in text
System.out.println(txt);

// Journal output.
   Min.   1st Qu.    Median     Mean    3rd Qu.     Max.
-3.19700  -0.65330  -0.09893  -0.07190   0.53300   2.29000
```

Install a new class library.

```
// Install new class library.
System.out.println(s.installPackage("sensitivity", true));

// Journal output.
  trying to load package sensitivity
  package sensitivity is not installed.
  package sensitivity not yet installed.
[eval] install.packages('sensitivity',repos='http://cran.cict.
fr/',dependencies=TRUE)
  org.rosuda.REngine.REXPNull@4d47c5fc
  request package sensitivity install...
  package sensitivity is not installed.
!   package sensitivity installation failed.
Impossible to install package sensitivity !
```

It turns out that Rsession is more user-friendly when compared to the JavaAPI of Rserve in Section 5.1. Rsession encapsulated the process of Java calling R, which will make it easier for those with a Java background to get started and master statistical calculation with Java application. Let's be creative!

4.3 High-Speed Channel between R and rJava

Question

Is there a two-way channel between Java and R?

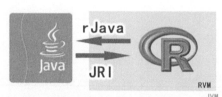

High-Speed Channel between R and Java

http://blog.fens.me/r-rjava-java/

Java has dominated the industry for quite a long time. Java syntax, JVM, JDK, and Java open source libraries have all gained explosive growth and covered almost all fields of application development. As Java covers more fields, problems also occur. It is becoming more and more difficult to learn Java as the syntax is getting more complex and similar projects are created every day. It is even harder for statistical practitioners without an IT background to learn to use Java.

R has always been an outstanding language in statistics as it has a simple syntax and moderate learning curve. It will be very useful to combine the universality of Java with the professionalism of R. This section will introduce the high-speed channel, rJava, to connect R and Java and realize two-way communication.

4.3.1 Introduction to RJava

RJava is a communication channel for R and Java. It allows R to directly call the objects and methods of Java through underlying JNI. RJava also provides the function for Java to call R, which is implemented through JRI(Java/R/Interface). JRI has now been implanted in the packages of rJava, and now we can try this function alone. RJava has now been a basic function component for many R packages based on Java.

RJava, being an underlying interface and using JNI as calling interface, is very efficient. In the program of JRI, JVM directly loads the calculation engine of R through memory, which means that the whole process of calling doesn't lose any performance. Thus rJava is a very efficient channel between R and Java and becomes the first priority for developers.

4.3.2 Installation of RJava

Linux environment used in this section:

■ Linux: Ubuntu 12.04.2 LTS 64bit
■ R: 3.0.1 x86_64-pc-linux-gnu
■ Java: Oracle SUN 1.6.0_29 64bit

Note: rJava supports both Windows 7 and Linux.

I suggest using root authority to install rJava. Because rJava is a basic package for cross-platform calling, we can reduce errors in the authority check of Linux when cross-platform programs are called by using root authority.

We assume that the Java environment is already installed before we install rJava. For the installation of the Java environment, please refer to Appendix A.

```
# Configure rJava environment using root authority.
~ sudo R CMD javareconf

# Start R using root authority.
~ sudo R

# Install rJava.
> install.packages("rJava")
installing via 'install.libs.R' to/usr/local/lib/R/site-library/rJava
** R
** inst
** preparing package for lazy loading
** help
*** installing help indices
** building package indices
** testing if installed package can be loaded
* DONE (rJava)

The downloaded source packages are in
        '/tmp/RtmpiZyCE7/downloaded_packages'

# Load rJava.
> library(rJava)
```

4.3.3 Implement R Calling Java Using RJava

Use rJava to program in the R environment.

```
# Check the loaded packages of current environment.
> search()
 [1] ".GlobalEnv"        "package:rJava"     "package:stats"
 [4] "package:graphics"  "package:grDevices" "package:utils"
 [7] "package:datasets"  "package:methods"   "Autoloads"
[10] "package:base"

# Start JVM.
> .jinit()

# Declare and assign to character string.
> s <- .jnew("java/lang/String", "Hello World!")
> s
[1] "Java-Object{Hello World!}"
```

```
# Check the length of character string.
> .jcall(s,"I","length")
[1] 12

# Index the position of "World".
> .jcall(s,"I","indexOf","World")
[1] 6

# View the method declaration of method "concat".
> .jmethods(s,"concat")
[1] "public java.lang.String java.lang.String.concat(java.lang.String)"

# Use method "concat" to connect character string.
> .jcall(s,"Ljava/lang/String;","concat",s)
[1] "Hello World!Hello World!"

# Print character string object.
> print(s)
[1] "Java-Object{Hello World!}"

# Print the value of character string.
> .jstrVal(s)
[1] "Hello World!"
```

Use $ to call method optimized by rJava.

```
# Same to jcall(s,"I","length").
> s$length()
[1] 12

# Same to.jcall(s,"I","indexOf","World").
> s$indexOf("World")
[1] 6
```

4.3.4 Implement Java Calling R Using RJava(JRI) (Windows 7)

Install rJava in Windows 7.

System environment used in this section:

- Windows 7: x86_64-w64-mingw32/x64 (64-bit)
- R: version 3.0.1
- Java: Oracle SUN JDK 1.6.0_45 64bit

Configure environment variables.

```
PATH: C:\Program Files\R\R-3.0.1\bin\x64;D:\toolkit\java\jdk6\bin;;
D:\toolkit\java\jdk6\jre\bin\server
JAVA_HOME: D:\toolkit\java\jdk6
CLASSPATH: C:\Program Files\R\R-3.0.1\library\rJava\jri
```

Install rJava in R.

```
# Start R.
~ R

# Install rJava.
> install.packages("rJava")

# Load rJava.
> library(rJava)
> .jinit()

# Test of R calling Java variables.
> s <-.jnew("java/lang/String", "Hello World!")
> s
[1] "Java-Object{Hello World!}"
```

Start Eclipse to write programs, as in Figure 4.5.
Then let's write a class of Java to finish the implementation of calling.

■ Method main(): the entrance of the start of program. Instantiate a demo object and call method callRJava().
■ Method callRJava(): create an Rengine object and start R engine in JVM. Transfer two statements of R to R engine in the form of character string and output the result in Java.

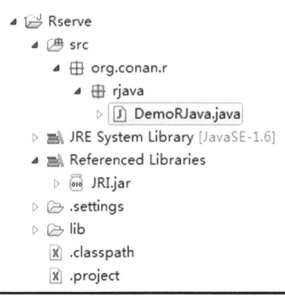

Figure 4.5   Create Rserve projects in Eclipse.

Create Java program file DemoRJava.java.

```java
package org.conan.r.rjava;

import org.rosuda.JRI.Rengine;

public class DemoRJava {

    /**
     * Method main for starting Java applications.
     */
    public static void main(String[] args) {
        DemoRJava demo = new DemoRJava();
        demo.callRJava();
    }

    public void callRJava() {
        Rengine re = new Rengine(new String[] {"—vanilla"}, false,
null);//Instantiate Rengine object and start R engine.
        if (!re.waitForR()) {
            System.out.println("Cannot load R");
            return;
        }
        //Execute a statement.
        String version = re.eval("R.version.string").asString();
        System.out.println(version);//Print the result.

        //Print the array repeatedly.
        double[] arr = re.eval("rnorm(10)").asDoubleArray();
        for (double a: arr) {
            System.out.print(a + ",");
        }
        re.end();//Close R engine.
    }
}
```

Configure startup parameters of VM in Eclipse, as in Figure 4.6.

```
Djava.library.path="C:\Program Files\R\R-3.0.1\library\rJava\jri\x64"
```

Run result:

```
R version 3.0.1 (2013-05-16)
0.04051018703700011,-0.3321596519938258,0.45642459001166913,
-1.1907153494936031,1.5872266854172385,1.3639721994863943,
-0.6309712627586983,-1.5226698569087498,-1.0416402147174952,
0.4864034017637044,
```

Package DemoRJava.jar in Eclipse and upload it to Linux to continue the test.

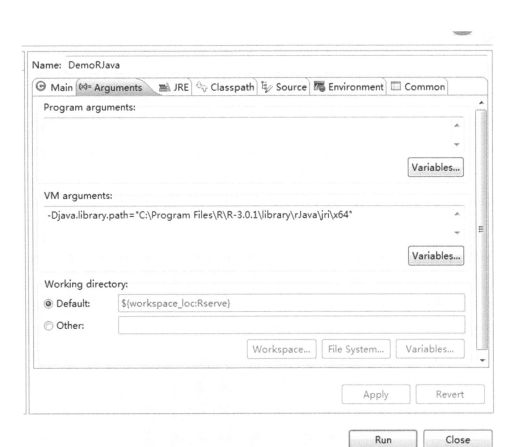

Figure 4.6 Configure the startup parameters of Eclipse.

4.3.5 Implement Java Calling R Using RJava(JRI) (Ubuntu)

Create a directory DemoRJava and upload DemoRJava.jar to DemoRJava. The following is the Shell command.

```
~ mkdir/home/conan/R/DemoRJava
~ cd/home/conan/R/DemoRJava
~ ls -l
-rw-r-r- 1 conan conan 1328 Aug 8 2013 DemoRJava.jar
```

Run the Jar package.

```
~ export R_HOME=/usr/lib/R
~ java -Djava.library.path=/usr/local/lib/R/site-library/rJava/jri
-cp/usr/local/lib/R/site-library/rJava/jri/JRI.jar:/home/conan/R/
DemoRJava/DemoRJava.jar org.conan.r.rjava.DemoRJava
```

Run the result.

```
R version 3.0.1 (2013-05-16)
0.6374494596732511,1.3413824702002808,0.04573045670001342,
-0.6885617932810327,0.14970067632722675,-0.3989493870007832,
-0.6148250252955993,0.40132038323714453,
-0.5385260423222166,0.3459850956295771,
```

Thus we've achieved two-way calling between R and Java using rJava and JRI in both Windows 7 and Linux Ubuntu.

4.4 Cross-Platform Communication between Node.js and R

Question

How do we call R in Node.js?

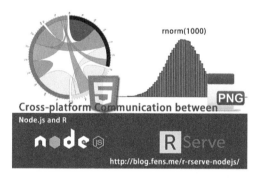

Programmers who do not use Nodejs in Web development may have fallen a little bit behind. Nodejs is a development platform of background programs based on JavaScript. In terms of my personal experience, it's more efficient in development than PHP. Though completely adopting asynchronous loading, Node.js still has great potential to surpass PHP in performance.

HTML5, as a Web front end that uses JavaScript a lot, has an abundance of beautiful effects. If we use HTML5 to re-render the images of R and add communication and user interaction parts to them, the results will be a combination of the advantages of both languages and surely will be amazing. The following is an introduction to cross-platform communication between R and Node.js.

4.4.1 Introduction to Node.js

Node.js is a platform based on Chrome's JavaScript runtime, which can be used to quickly construct Web services and applications. In other words, Node,js encapsulates the V8 engine of Google (used on Google Chrome). The speed of the V8 engine in running JavaScript is very fast, and the performance is also very good. Node.js optimizes some special cases and provides alternative API, which ensures a good performance of V8 in a non-browser environment.

4.4.2 Environment Configuration of R

For the cross-platform communication between R and Node.js in this section, the support library of R is Rserve. The communication is achieved through a TCP/IP communication protocol of R provided by Rserve. I've already finished the configuration of the Rserve environment of Linux Ubuntu. The system environment used in this section is as follows:

- Linux: Ubuntu 12.04.2 LTS 64bit
- R: 3.0.1 x86_64-pc-linux-gnu
- IP: 192.168.1.201

Rserve environment: Rserve v1.7-1. Interface: 6311, remote access allowed. View the process of Rserve:

```
~ ps -aux|grep Rserve
conan     9736   0.0  1.2 116288 25440 ?         Ss   13:11   0:01/
usr/lib/R/bin/Rserve --RS-enable-remote

~ netstat -nltp|grep Rserve
tcp        0      0 0.0.0.0:6311
0.0.0.0:*              LISTEN        9736/Rserve
```

4.4.3 Environment Configuration of Node.js

The Node.js environment is a Web framework based on Express3. Library rio is the dependent library of the communication between Node.js and Rserve. If you are a beginner in Node.js and plan to learn it from the very beginning, please refer to the author's blog, *Learning Node.js from the very beginning* series. If you only need to build a test environment of Node.js, please refer to the author's article *Preparing the Development Environment Ubuntu of Node.js*.

Development environment of Windows 7:

- Node: v0.10.5
- NPM: 1.2.19
- IP: 192.168.1.13
- Express: 3.2.2

We've created Express3 project in Windows 7. The following is the project directory: D:\workspace\project\investment\webui.

```
~ D:\workspace\project\investment\webui>ls
README.md  app.js  models  node-rio-dump.bin  node_modules  package.
json  public  routes  views
```

Install library rio.

```
~ D:\workspace\project\investment\webui>npm install rio
rio@0.9.0 node_modules\rio
├── hexy@0.2.5
└── binary@0.3.0 (buffers@0.1.1, chainsaw@0.1.0)
```

4.4.4 Cross-Platform Communication between Node.js and R

Now let's write Node.js program to achieve cross-platform communication. First, add a routing path in the configuration file app.js.

```
~ vi app.js
// Omit.
var vis = require('./routes/vis')
app.get('/vis/rio',vis.rio);
```

Add a routing file in the routing directory/routes.

```
~ vi /routes/vis.js

var rio = require("rio");
exports.rio = function(req, res){
  options = {
     //Address of remote Rserve server.
     host: "192.168.1.201",
     //Interface of remote Rserve server.
     port : 6311,
  // Callback function of computed result.
  callback: function (err, val) {
       // Normal return.
       if (!err) {
         console.log("RETURN:"+val);
         // Transfer the result to interface.
         return res.send({'success':true,'res':val});

          } else {    // Error return.
             console.log("ERROR:Rserve call failed")
             return res.send({'success':false});
          }
       },
    }
  rio.enableDebug(true);                      //Start debug mode.
  rio.evaluate("pi/2 * 2 * 2",options);    //Run R codes.
};
```

Through the preceding codes, we achieved the remote connection of rio and Rserve. Transfer the statement (pi/2*2*2), as a parameter, to a remote Rserve server in the form of characters, and get the return value through the callback method.

Open a browser(http://localhost:3000/vis/rio). It can be seen from the Web interface that the computed result of (pi/2*2*2) is 6.283185307179586. The structure return value in the browser is a JSON object.

```
{
  "success": true,
  "res": 6.283185307179586
}
```

Output the command line log.

```
Connected to Rserve
Supported capabilities --------------

Sending command to Rserve
00000000: 0300 0000 1400 0000 0000 0000 0000 0000  ................
00000010: 0410 0000 7069 202f 2032 202a 2032 202a  ....pi./.2.*.2.*
00000020: 2032 0001                                .2..

Data packet
00000000: 2108 0000 182d 4454 fb21 1940            !....-DT{!.@

Type SEXP 33
Response value: 6.283185307179586
RETURN:6.283185307179586
GET/vis/rio 200 33ms - 49b
Disconnected from Rserve
Closed from Rserve
```

I can see the communication situation between Node.js and Rserve: the response value is 6.283185307179586, which is the same with the display on the page. Then let's modify the running script of R. Rnorm(10) is to take 10 random numbers of standard normal distribution, N(0,1).

```
rio.evaluate("rnorm(10)",options);//Run R codes.
```

The JSON object of the computed result returned by the browser.

```
{
  "success": true,
  "res": [
    -0.011531884725262991,
    0.5106443501593562,
    -0.05216533321965309,
```

```
    1.9221980152236238,
    0.5205238122633465,
    -0.3275367539102907,
    -0.06588102930129405,
    1.5410418730008988,
    1.308169913050071,
    0.005044179478212583
  ]
}
```

Command line log.

```
Connected to Rserve
Supported capabilities --------------

Sending command to Rserve
00000000: 0300 0000 1000 0000 0000 0000 0000 0000  ................
00000010: 040c 0000 726e 6f72 6d28 3130 2900 0101  ....rnorm(10)...

Data packet
00000000: 2150 0000 f6ca 0c5e 079e 87bf 9b4a fad1  !P..vJ.^...?.JzQ
00000010: 3257 e03f eda2 5320 6ab5 aabf 2b25 bdb4  2W`?m"S.j5*?+%=4
00000020: 52c1 fe3f ebba ce8d 21a8 e03f bc17 92b7  RA~?k:N.!(`?<..7
00000030: 5cf6 d4bf ca9f 4642 94dd b0bf 1be3 e485  \vT?J.FB.]0?.cd.
00000040: 1ba8 f83f 5a94 2293 43ee f43f 1724 4e9e  .(x?Z.".Cnt?.$N.
00000050: 34a9 743f                                4)t?

Type SEXP 33
Response value: -0.011531884725262991,0.5106443501593562,
-0.05216533321965309,1.9221980152236238,0.5205238122633465,
-0.3275367539102907,-0.06588102930129405,1.5410418730008988,
1.308169913050071,0.005044179478212583
RETURN:-0.011531884725262991,0.5106443501593562,
-0.05216533321965309,1.9221980152236238,0.5205238122633465,
-0.3275367539102907,-0.06588102930129405,1.5410418730008988,
1.308169913050071,0.005044179478212583
GET/vis/rio 200 30ms - 285b
Disconnected from Rserve
Closed from Rserve
```

Thus we've achieved the cross-platform communication between R and Node.js, which seems similar to the concept of "getting to a new level" in Chinese Kung-fu.

Chapter 5

Server Implementation of R

This chapter mainly introduces four tool packages of the R server, which may help readers to use R to create the running environment of Socket server, Web server, and WebSocket server.

5.1 A Detailed Elaboration of the Server Program of R: Rserve

Question

How do we use Rserve?

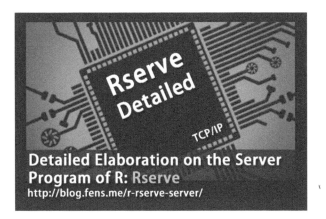

We've met Rserve for the connection between R and Java in Section 4.1. Now let's learn more details of Rserve.

Many projects depend on Rserve, as a communication Transmission Control Protocol/Internet Protocol (TCP/IP) interface between R and many other languages. The server configuration and operation of Rserve is very easy to learn, and its client is implemented by many languages including C/C++, Java, and so forth. R has its own client to implement the RSclient project, which will be introduced in the next section. This section provides a detailed discussion of the configuration and use of Rserve as a server application.

5.1.1 Start of Rserve

System environment used in this section:

- Linux: Ubuntu 12.04.2 LTS 64bit
- R: 3.0.1 x86_64-pc-linux-gnu

Note: Rserve supports both Windows 7 and Linux. Because Rserve is mainly used as a communication server, Linux is recommended more.

The following is the installation of Rserve:

```
# Start R.
~ R

# Install Rserve.
> install.packages("Rserve")

# Load Rserve.
> library(Rserve)
```

There are two ways to start the Rserve server: starting the Rserve server in R and starting the Rserve server in the command line. To start the Rserve server in R, we may need the functions of Rserve.

- Rserve(): start a daemon process as Rserve instance alone
- run.Rserve(): start Rserve instance in the current process
- self(): mainly used for the program interaction with the Rserve server in the current process, including four functions: self.ctrlEval(), self.ctrlSource(), self.oobSend(), self.oobMessage()

5.1.1.1 Start the Rserve Server in the Program

```
> library(Rserve)

# Start a daemon process as Rserve instance alone.
> Rserve()
Starting Rserve:
 /usr/lib/R/bin/R CMD/home/conan/R/x86_64-pc-linux-gnu-library/3.0/
Rserve/libs//Rserve
```

View the Rserve process.

```
~ ps aux | grep R
conan     8799  0.1  1.5 121748 32088 pts/0     S+   22:30   0:00
/usr/lib/R/bin/exec/R
conan     8830  0.0  1.2 116336 25044 ?         Ss   06:46   0:00
/home/conan/R/x86_64-pc-linux-gnu-library/3.0/Rserve/libs//Rserve
```

```
~ netstat -nltp|grep Rserve
tcp        0      0 127.0.0.1:6311           0.0.0.0:*
LISTEN       8830/Rserve
```

In this circumstance, the current R environment is not interrupted. Instead, an Rserve instance is started alone in the system background, and using run.Rserve() is to start Rserve in the current R environment.

```
# Start Rserve instance in the current process.
> run.Rserve()
-- running Rserve in this R session (pid=30664), 1 server(s) --
(This session will block until Rserve is shut down)

# View the R process.
~ ps aux|grep R
30664 pts/0    00:00:00 R

~ netstat -nltp|grep R
tcp        0      0 127.0.0.1:6311           0.0.0.0:*
LISTEN       30664/R
```

5.1.1.2 Start the Rserve Server in the Command Line

First, view the command line help of Rserve.

```
~ R CMD Rserve --help
Usage: R CMD Rserve []

Options: --help  this help screen
 --version  prints Rserve version (also passed to R)
 --RS-port    listen on the specified TCP port
 --RS-socket   use specified local (unix) socket instead of TCP/IP.
 --RS-workdir   use specified working directory root for connections.
 --RS-encoding   set default server string encoding to.
 --RS-conf   load additional config file.
 --RS-settings   dumps current settings of the Rserve
 --RS-source    source the specified file on startup.
 --RS-enable-control   enable control commands
 --RS-enable-remote   enable remote connections

All other options are passed to the R engine.
```

Start Rserve in the command line and open the remote access mode.

```
~ R CMD Rserve --RS-enable-remote

R version 3.0.1 (2013-05-16) — "Good Sport"
```

```
Copyright (C) 2013 The R Foundation for Statistical Computing
Platform: x86_64-pc-linux-gnu (64-bit)

R is free software and comes with ABSOLUTELY NO WARRANTY.
You are welcome to redistribute it under certain conditions.
Type 'license()' or 'licence()' for distribution details.

  Natural language support but running in an English locale

R is a collaborative project with many contributors.
Type 'contributors()' for more information and
'citation()' on how to cite R or R packages in publications.

Type 'demo()' for some demos, 'help()' for on-line help, or
'help.start()' for an HTML browser interface to help.
Type 'q()' to quit R.

Rserv started in daemon mode.
```

View the Rserve process.

```
~ ps -aux|grep Rserve
conan    27639  0.0  1.2 116288 25236 ?         Ss   20:41   0:00
/usr/lib/R/bin/Rserve --RS-enable-remote

~ netstat -nltp|grep Rserve
tcp       0      0 0.0.0.0:6311              0.0.0.0:*
LISTEN      27639/Rserve
```

5.1.2 Advanced Use of Rserve: Rserve Configuration Management

We can manage the Rserve server and define the starting script of server through the configuration file Rserv.conf. View the default configuration information of Rserve server through the following command.

```
~ R CMD Rserve --RS-settings
Rserve v1.7-1

config file:/etc/Rserv.conf
working root: /tmp/Rserv
port: 6311
local socket: [none, TCP/IP used]
authorization required: no
plain text password: not allowedv
passwords file: [none]
allow I/O: yes
allow remote access: no
control commands: no
interactive: yes
max.input buffer size: 262144 kB
```

Configuration instruction of the current Rserve server:

- config file: system will skip this item if there is no/etc/Rserve.conf in local file
- working root: working directory/tmp/Rserv when R is run
- port: communication port 6311
- local socket: TCP/IP protocol
- authorization: authorization is not started
- plain text password: plain text password is not allowed
- password file: password file, not specified
- allow I/O: allow IO operation
- allow remote access: remote access is not started
- control commands: control command is not started
- interactive: communication is allowed
- max.input buffer size: uploading file size limits, 262 mb

Modify default configuration, add remote access and set loading script. Create new file /etc/Rserv.conf.

```
~ sudo vi /etc/Rserv.conf

workdir /tmp/Rserv
remote enable
fileio enable
interactive yes
port 6311
maxinbuf 262144
encoding utf8
control enable
source /home/conan/R/RServe/source.R
eval xx=1
```

The option "source" is used for configuring the loaded file when the Rserve server is started, including initializing system variables and system functions, and so forth. The option "eval" is used to define environment variables.

Add the initialized starting script of the Rserve server.

```
~ vi /home/conan/R/RServe/source.R

cat("This is my Rserve!!")
print(paste("Server start at",Sys.time()))
```

View the server configuration again.

```
~ R CMD Rserve --RS-settings
Rserve v1.7-1
config file: /etc/Rserv.conf
working root: /tmp/Rserv
port: 6311
```

```
local socket: [none, TCP/IP used]
authorization required: yes
plain text password: allowed
passwords file: [none]
allow I/O: yes
allow remote access: yes
control commands: yes
interactive: yes
max.input buffer size: 262144 kB
```

Restart the Rserve server.

```
~  R CMD Rserve
R version 3.0.1 (2013-05-16) -- "Good Sport"
Copyright (C) 2013 The R Foundation for Statistical Computing
Platform: x86_64-pc-linux-gnu (64-bit)

R is free software and comes with ABSOLUTELY NO WARRANTY.
You are welcome to redistribute it under certain conditions.
Type 'license()' or 'licence()' for distribution details.

  Natural language support but running in an English locale

R is a collaborative project with many contributors.
Type 'contributors()' for more information and
'citation()' on how to cite R or R packages in publications.

Type 'demo()' for some demos, 'help()' for on-line help, or
'help.start()' for an HTML browser interface to help.
Type 'q()' to quit R.

This is my Rserve!![1] "Server start at 2013-10-30 22:38:10"
Rserv started in daemon mode.
```

View the log: source.R is executed when started.

```
"This is my Rserve!![1] "Server start at 2013-10-30 22:38:10""
```

View the process.

```
~ ps -aux|grep Rserve
conan    28339   0.0  1.2 116292 25240 ?        Ss    22:31   0:00/
usr/lib/R/bin/Rserve

~ netstat -ntlp|grep Rserve
tcp      0      0 0.0.0.0:6311            0.0.0.0:*
LISTEN      28339/Rserve
```

0.0.0.0 indicates that remote access is allowed and IP is not restricted.

5.1.3 Advanced Use of Rserve: Users' Login Authentication

Use RSclient to access Rserve in the current environment without authentication. For the use of RSclient, please refer to Section 5.2

```
~ R
> library(RSclient)  # Load RSclient.
> conn<-RS.connect()
> RS.eval(conn,rnorm(10))
 [1]  0.03230305  0.95710725 -0.33416069 -0.37440009 -1.95515719
-0.22895924
 [7]  0.39591984  1.67898842 -0.01666688 -0.26877775
```

Modify the configuration, add users' login authentication, and permit plain text password. Modify the file /etc/Rserv.conf.

```
~ sudo vi /etc/Rserv.conf

workdir /tmp/Rserv
remote enable
fileio enable
interactive yes
port 6311
maxinbuf 262144
encoding utf8
control enable
source /home/conan/R/RServe/source.R
eval xx=1
auth required
plaintext enable
```

Authentication reports an error when using RSclient again to access Rserve direct.

```
> library(RSclient)
> conn<-RS.connect()
> RS.eval(conn,rnorm(10))
Error in RS.eval(conn, rnorm(10)) :
  command failed with status code 0x41: authentication failed
```

Use RSclient to log in and connect.

```
> library(RSclient)
> conn<-RS.connect()
> RS.login(conn,"conan","conan",authkey=RS.authkey(conn))
[1] TRUE
> RS.eval(conn,rnorm(5))
[1] -1.19827684  0.72164617  0.22225934  0.09901505 -1.54661436
```

The users' login authentication here is the binding operating system users. We could also specify the uid and gid parameter in Rserve.conf to control the server permissions on a more detailed level.

This section gives a detailed introduction to the installation, start, configuration, and use of Rserve. With this knowledge, we can now use Rserve to construct enterprise online applications.

5.2 Client of Rserve in R: RSclient

Question

How do we access Rserve using native R?

RSclient is a client program of R to implement the communication of Rserve. Such communication is of great help to the real application structure, as it will not only unify the interface of Rserve but also implement the cross virtual machine step-by-step program design of R.

We've previously discussed that Rserve is a network communication server program based on TCP/IP protocol between R and other languages based on a Client/Server (C/S) structure. So how do we implement the communication between R and Rserve server? RSclient is the solution to this question and makes it possible for R to access the Rserve server instance. Thus R gains its basis of step-by-step program calling.

5.2.1 Configuration of the Rserve Server

System environment used in this section:

- Linux: Ubuntu 12.04.2 LTS 64bit
- R: 3.0.1 x86_64-pc-linux-gnu
- Rserve: Rserve v1.7-1

Note: Rserve supports both Windows 7 and Linux. Because Rserve is mainly used as a communication server, Linux is recommended more.

Start the Rserve server.

```
# Start Rserve through command line.
~ R CMD Rserve

~ ps -aux|grep Rserve
conan    28339  0.0  1.2 116292 25240 ?          Ss   22:31   0:00
/usr/lib/R/bin/Rserve

~ netstat -ntlp|grep Rserve
tcp        0      0 0.0.0.0:6311            0.0.0.0:*
LISTEN       28339/Rserve
```

Rserve environment:

◼ IP: 192.168.1.201, allows remote access
◼ Port: 6311
◼ Login authentication: username: conan, password: conan
◼ Character encoding: utf-8

```
# View the configuration of Rserve server.
~ R CMD Rserve --RS-settings
Rserve v1.7-1
config file: /etc/Rserv.conf
working root: /tmp/Rserv
port: 6311
local socket: [none, TCP/IP used]
authorization required: yes
plain text password: allowed
passwords file: [none]
allow I/O: yes
allow remote access: yes
control commands: yes
interactive: yes
max.input buffer size: 262144 kB
```

After Rserve is started, we'll use RSclient to access the Rserve server.

5.2.2 Installation of RSclient

Because the communication between RSclient and Rserve can be done through remote access, we use Windows 7 to install RSclient. System environment of RSclient:

◼ Windows 7 64bit
◼ R: 3.0.1 x86_64-w64-mingw32/x64 b4bit

Installation and loading of RSclient.

```
# Start R.
~ R
```

```
# Install RSclient.
> install.packages("RSclient")

# Load RSclient.
> library(RSclient)
```

5.2.3 API of RSclient

The API of RSclient can be divided into two groups: Rclient (the old version) and RCC (the new version). The API function name of Rclient is as follows.

```
RSassign        RSattach        RSclose
RSconnect       RSdetach        RSeval          RSevalDetach
RShowDoc        RSiteSearch     RSlogin         RSserverEval
RSserverSource  RSshutdown      RSclient
```

API function name of the new version RCC is splited by ".", as follows.

```
RS.assign        RS.authkey       RS.close         RS.collect
RS.connect       RS.eval          RS.eval.qap      RS.login
RS.oobCallbacks  RS.server.eval   RS.server.shutdown RS.server.
source
RS.switch
```

This section mainly introduces the use of API in the new version. The folllowing is the operating function of client:

- RS.connect: create the connection with Rserve
- RS.close: close the connection with Rserve
- RS.login: login authentication
- RS.authkey: set the encryption algorithm in authentication
- RS.eval: run R statements remotely in Rserve
- RS.eval.qap: run R statements remotely and use serialized Rserve QAP objects to replace local objects
- RS.collect: wait for the asynchronous execution eval result and return
- RS.assign: execute assignment remotely
- RS.oobCallbacks: callback function, executed through OOB_SEND function and OOB_MSG function

Server management functions (need to set −RS-enable-control when Rserve is started) are as follows:

- RS.server.eval: server control function, execute script
- RS.server.shutdown: server control function, close server
- RS.server.source: server control function, execute local file in the server

5.2.4 Use of RSclient

```
> library(RSclient)

# Create remote connection
> conn<-RS.connect(host="192.168.1.201")
> conn
 Rserve QAP1 connection 0x000000000445cd60 (socket 308, queue length
0)

# Sign in Rserve server and finish authentication check.
> RS.login(conn,"conan","conan",authkey=RS.authkey(conn))
[1] TRUE

# Run script.
> RS.eval(conn,rnorm(5))
[1] -2.6762608  1.4435144 -0.4298395 -0.7046573 -1.4056073

# Set variables.
> RS.assign(conn,"xx",99)
raw(0)
> RS.eval(conn,xx-55)
[1] 44

# Synchronous execution.
> RS.eval(conn,head(rnorm(10000000)),wait=TRUE)
[1] -4.20217390  0.22353317 -1.70256992  0.30053213 -0.01427486
-0.70522254

# Asynchronous execution. The result should be taken again through
RS.collect().
> RS.eval(conn,head(rnorm(10000000)),wait=FALSE)
NULL
> RS.collect(conn)
[1] -0.2814752  0.3215521 -1.0978825 -0.8534461 -0.2459560
-0.4804882

# Close connection.
> RS.close(conn)
NULL
> conn
 Closed Rserve connection 0x000000000445cc80
```

5.2.5 Simultaneous Access of Two Clients

Operation of client A.

```
~ R
> library(RSclient)
> conn<-RS.connect(host="192.168.1.201")
```

```
> RS.login(conn,"conan","conan",authkey=RS.authkey(conn))
> RS.assign(conn,"A",1234)

> RS.eval(conn,A)
[1] 1234

> RS.eval(conn,getwd())
[1] "/tmp/Rserv/conn29039"
```

Operation of client B.

```
~ R
> library(RSclient)
> conn<-RS.connect(host="192.168.1.201")
> RS.login(conn,"conan","conan",authkey=RS.authkey(conn))
> RS.assign(conn,"B",5678)

> RS.eval(conn,B)
[1] 5678
> RS.eval(conn,ls())
[1] "B"
> RS.eval(conn,getwd())
[1] "/tmp/Rserv/conn29040"
```

We can see that after the ports of client A and client B are created, Rserve on the server side will be operated in two separate spaces. Therefore the accesses of client A and client B are separate. So how could we test the communication between two clients? The answer is through global variables. Set global variables in Rserve server. The code is as follows.

```
> RS.server.eval(conn, "G<-999"):
command failed with status code 0x48: access denied
```

Error of "access rejected" is reported. Though we have opened –RS-enable-control, there is still an error. We don't know whether it's a bug of Rserve, but we would know that global variables have proved to be impracticable. Then we could use intermediate variables to achieve an interaction between two clients indirectly and to store the data needed to be interacted into MySQL and Redis.

We've achieved the remote connection between Rserve server and R through RSclient. If we expand our thought further, we could construct a step-by-step computing environment through Rserve and RSclient. With other open source technique as assistance, R can also reach the standard of an online application.

5.3 An R Program Running on the Web: FastRWeb

Question

How do we display the image made by R on the Web?

R has long been used in a client program based on personal computers. We are familiar with the whole process: downloading installation package, installing on desktop, writing algorithm, and running the codes. Then we publish our work in the form of either images or documents. But if we can run R on the server side and publish the result on the Web, we will be working in an Internet style! FastRWeb provides us a way to implement an R application of a Browser/Server (B/S) structure.

5.3.1 Introduction to FastRWeb

FastRWeb is a basic architecture environment. It allows R script to run on any Web servers to display data and images. Users can communicate and interact through URL addresses and R script. FastRWeb can construct a Web environment of R very fast. FastRWeb is based on a CGI program, which means that all Web servers supporting a CGI program can run FastRWeb. Now, I'll take Rserve as an example to deploy FastRWeb and implement Web application of R.

The architecture of R can be seen in Figure 5.1, where we can find the architecture principle. (1) Browser accesses the Web server through an http request. (2) The Web server sends the request to the Rserver server through a socket. (3) Rserve calls FastRWeb, runs R script, and returns data and images. (4) Browser gets the result and displays it on the Web.

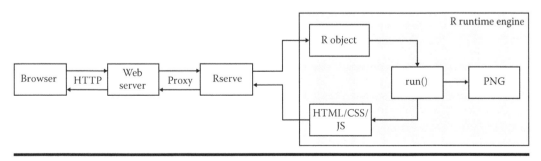

Figure 5.1 Architecture of FastRWeb.

5.3.2 Installation of FastRWeb

System environment used in this section:

- Linux: Ubuntu 12.04.2 LTS 64bit
- R: 3.0.1 x86_64-pc-linux-gnu
- IP: 192.168.1.201

Note: FastRWeb only supports a Linux environment.
Installation of FastRWeb is as follows.

```
~ R
> install.packages("FastRWeb")
```

Because FastRWeb depends on Cairo, Cairo will be installed on the local library of Linux. Please refer to Section 1.7 for this part. Now let's install Rserve. For the installation and use of Rserve, please refer to Section 5.1.

```
> install.packages("Rserve")
```

Create Rserve environment of FastRWeb based on Rserve.

```
# Enter the installation directory of FastRWeb.
~ cd/home/conan/R/x86_64-pc-linux-gnu-library/3.0/FastRWeb

# Run installation script.
~ sudo ./install.sh
Done.
Please check files in/var/FastRWeb/code
If they match tour setup, you can start Rserve using
/var/FastRWeb/code/start
```

View the generated directory,/var/FastRWeb/code/.

```
~ cd/var/FastRWeb/code/

~ ls -l
-rw-r--r-- 1 conan conan 1210 Oct 29 16:07 README
-rw-r--r-- 1 conan conan   79 Oct 29 17:50 rserve.conf
-rw-r--r-- 1 conan conan 2169 Oct 29 17:51 rserve.R
-rwxr-xr-x 1 conan conan  457 Oct 29 17:35 start
```

README is the help file, rserve.conf is the starting parameter of Rserve, rserve.R is the starting script of Rserve, and start is the command.

Modify the configuration file rserve.conf of starting Rserve.

```
~ vi rserve.conf

http.port 8888
remote enable
source /var/FastRWeb/code/rserve.R
control enable
```

By default, Rserve provides socket communication interface. For the convenience of the Web test, we'll use an http communication interface instead. Modify the file rserve.R and add two lines of codes to the top.

```
~ vi rserve.R

# Content added.
library(FastRWeb)
.http.request <- FastRWeb:::.http.request
```

Now we've completed the modification to use http protocol as the communication interface.

5.3.3 Use of FastRWeb

Start the FastRWeb service.

```
~ sudo ./start
R CMD Rserve --RS-conf/var/FastRWeb/code/rserve.conf --vanilla
--no-save
--RS-enable-remote

R version 3.0.1 (2013-05-16) -- "Good Sport"
Copyright (C) 2013 The R Foundation for Statistical Computing
Platform: x86_64-pc-linux-gnu (64-bit)

Starting Rserve on conan
Loading packages...
XML: TRUE
Cairo: TRUE
Matrix: TRUE
FastRWeb: TRUE
Rserv started in daemon mode.
```

We can see from the startup log that XML, Cairo, Matrix, FastRWeb, and Rserve have all been loaded and run normally. View the system process and port:

```
~ ps -aux|grep Rserve
conan     23739  0.0  1.4 120140 28916 ?          Ss   16:47   0:00

/usr/lib/R/bin/Rserve --RS-conf/var/FastRWeb/code/rserve.conf
--vanilla --no-save --RS-enable-remote

~ sudo netstat -nltp|grep Rserve
tcp        0       0 0.0.0.0:8888              0.0.0.0:*
LISTEN        25778/Rserve
tcp        0       0 0.0.0.0:6311              0.0.0.0:*
LISTEN        25778/Rserve
```

Two ports have been opened: one is the socket port of Rserve 6311, and the other is the http port 8888. Open http://192.168.1.201:8888/example1.png with browser and access through Web, as in Figure 5.2.

The corresponding file of R code of Figure 5.2 is/var/FastRWeb/web.R/example1.png.R.

```
~ vi/var/FastRWeb/web.R/example1.png.R

run <- function(...)  {
  p <- WebPlot(600, 600)
  plot(rnorm(100), rnorm(100), pch = 19, col = 2)
  p
}
```

Modify the file var/FastRWeb/web.R/example1.png.R and refresh in the browser. The result is in Figure 5.3.

```
~ vi/var/FastRWeb/web.R/example1.png.R

run <- function(...)  {
  p <- WebPlot(600, 600)
  plot(rnorm(400), rnorm(400), pch = 20, col = 3)
  p
}
```

Thus we've achieved running R script through the Web and drawing graphics on the Web. There are some other examples in the directory/var/FastRWeb/web.R/. You can take them as references and modify your own R script.

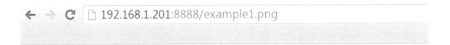

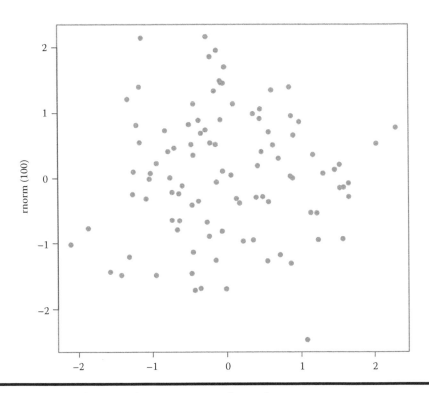

Figure 5.2   Visualized image of rnorm(100) on the Web.

```
~ ls -l/var/FastRWeb/web.R
total 32
-rw-r--r-- 1 conan conan 790 Oct 29 16:07 common.R
-rw-r--r-- 1 conan conan 316 Oct 29 20:01 example1.png.R
-rw-r--r-- 1 conan conan 520 Oct 29 16:07 example2.R
-rw-r--r-- 1 conan conan 174 Oct 29 16:07 index.R
-rw-r--r-- 1 conan conan 215 Oct 29 16:07 info.R
-rw-r--r-- 1 conan conan 64  Oct 29 16:07 main.R
-rw-r--r-- 1 conan conan 167 Oct 29 16:07 README
-rw-r--r-- 1 conan conan 214 Oct 29 16:07 tmp.R
```

The R script file example2.R allows transferring parameters on pages.

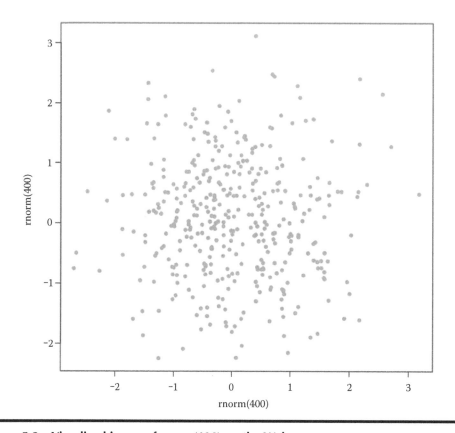

Figure 5.3   Visualized image of rnorm(400) on the Web.

According to the description of FastRWeb, it can communicate with any WebServer through CGI. In this way, we can employ R script on servers of advanced languages such as PHP, Python, Ruby, Java, and so forth. There is an article on R-bloggers discussing how to employ FastRWeb on XAMPP based on Apache (http://www.r-bloggers.com/setting-up-fastrweb-on-mac-os-x/). If you are interested in this, give it a try!

5.4 Building a WebSocket Server by R

Question
How do we implement WebSocket by R?

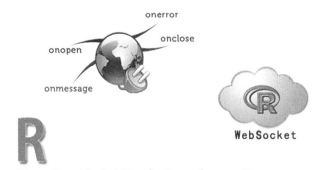

To Build WebSocket Server
http://blog.fens.me/r-websocket-websockets/

R has developed from a statistical language to an industrialized language. It not only supports the basic operation and visualization of the Web, but it also supports WebSocket. Our Internet application now can interact with R directly through the WebSocket protocol without using Rserve. R has undergone a technological revolution and become more advanced and convenient.

5.4.1 Introduction to WebSocket

WebSocket is a protocol based on HTML5 to implement the communication between clients and servers on browsers. WebSocket has four main advantages: (1) it decreases network overhead greatly; (2) it reduces the processing overhead of servers; (3) it simplifies the rapid asynchronous update of the Web client; and (4) it also simplifies the coupling status between servers and clients.

websockets is a class library of the WebSocket port of R. Through websockets, we can easily use R to construct a WebSocket server instance. The following is an introduction to API of websockets.

- create_server: create a WebSocket server instance and bind port
- daemonize: bond the daemon process of the WebSocket server instance to the console of R; it does not support Windows
- http_response: send http response request to socket
- http_vars: parse the parameter list of http GET/POST
- service: register the service queue of WebSocket instance
- set_callback: define R functions in WebSocket instance
- static_file_service: static file
- static_text_service: static text
- webSocket: create a WebSocket client instance
- webSocket_broadcast: issue broadcast to all clients registered in the same WebSocket server instance
- webSocket_close: close client connection
- webSocket_write: transfer data through WebSocket

5.4.2 Installation of websockets

System environment used in this section:

- Linux: Ubuntu Server 12.04.2 LTS 64bit
- R: 3.0.1 x86_64-pc-linux-gnu
- IP: 192.168.1.201

Note: websockets only supports Linux.
Installation of websockets.

```
# Start R.
~ R

# Install websockets.
> install.packages("websockets")

# Load websockets.
> library(websockets)
'websockets'R3.0.2
```

websockets depends on caTools, which is a tool set. Please refer to Section 1.8 for more information about CaTools.

The author found on June, 2015 that websockets had been from the CRAN library on March 2, 2014 and taken over and maintained again by Joe Cheng from RStudio. The address is http://cran.r-project.org/web/packages/websockets/index.html.

```
Package 'websockets' was removed from the CRAN repository.

Formerly available versions can be obtained from the archive.

Archived on 2014-03-02 at the request of the maintainer.
```

Thus when we install websockets, functions through install.packages() will report an error.

```
> install.packages("websockets")
Installing package into '/home/conan/R/
x86_64-pc-linux-gnu-library/3.0'
(as 'lib' is unspecified)
Warning:
package 'websockets' is not available (for R version 3.0.1)
```

We need to download the installation package and install it manually.

```
# Download the latest websockets.
~ wget http://cran.r-project.org/src/contrib/Archive/websockets/
websockets_1.1.7.tar.gz

# Install websockets in the current directory.
~ R CMD INSTALL websockets_1.1.7.tar.gz
* installing to library '/home/conan/R/
x86_64-pc-linux-gnu-library/3.0'
ERROR: dependencies 'caTools', 'digest' are not available for
package 'websockets'
* removing '/home/conan/R/x86_64-pc-linux-gnu-library/3.0/websockets'
```

Errors are reported during the installation. It warns of a lack of the dependent packages caTools and digest, so we need to install these two packages first.

```
# Start R.
~ R

# Install dependent packages.
> install.packages("caTools")
> install.packages("digest")

# Return to command line and re-install websockets. Succeeded.
~ R CMD INSTALL websockets_1.1.7.tar.gz
* installing to library '/home/conan/R/
x86_64-pc-linux-gnu-library/3.0'
* installing *source* package 'websockets'...
** Unpack websockets package successfully, md5 encryption and check.
** libs
gcc -std=gnu99 -I/usr/share/R/include -DNDEBUG    -DLWS_NO_FORK
-fpic  -O3 -pipe  -g  -c libsock.c -o libsock.o
gcc -std=gnu99 -shared -o websockets.so libsock.o -L/usr/lib/R/lib
-lR
installing to/home/conan/R/x86_64-pc-linux-gnu-library/3.0/
websockets/libs
** R
** demo
** inst
** preparing package for lazy loading
** help
*** installing help indices
** building package indices
** installing vignettes
'websockets.Rnw'
** testing if installed package can be loaded
* DONE (websockets)
```

```
# Start R.
~ R

# Load websockets.
> library(websockets)
```

Now we've installed websockets manually.

5.4.3 Quickly Start the WebSockets Server Demo

websockets provides a demo. Through demo(websockets), we can start a simple WebSocket server directly.

```
> library(websockets)
'websockets'R3.0.2

# Start demo.
> demo(websockets)
```

View the process of the server.

```
~ netstat -nltp|grep r

Proto Recv-Q Send-Q Local Address          Foreign Address
State       PID/Program name
tcp     0      0 0.0.0.0:7681            0.0.0.0:*
LISTEN     2231/rsession
```

If you open the page http://192.168.1.201:7681 in your browser and you can see the demo application implemented by websockets, as in Figure 5.4. Please note that the browser must support HTML5, so we recommend Chrome.

Output the log of the server:

```
Websocket client socket  20 has closed.
Websocket client socket  8 has been established.
Websocket client socket  21 has closed.
```

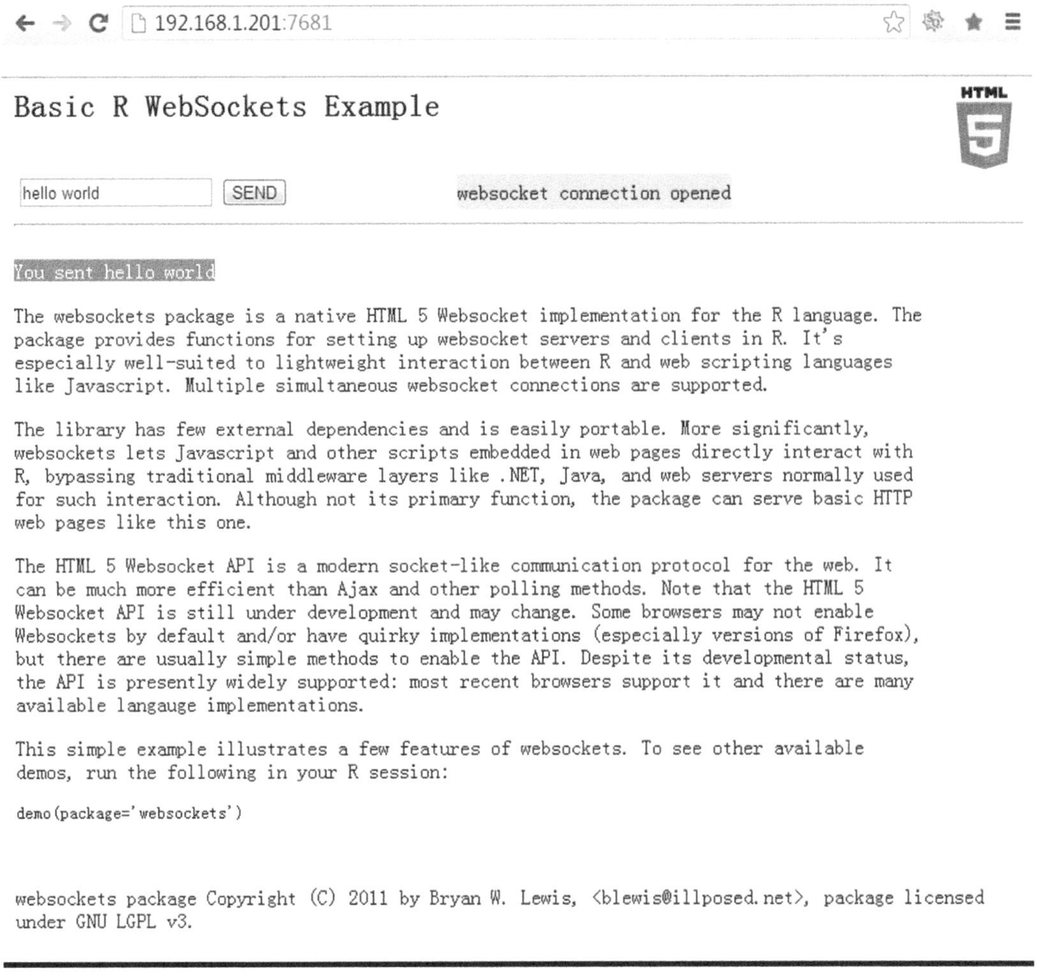

← → C 🗋 192.168.1.201:7681   ☆ 🕸 ★ ≡

Basic R WebSockets Example   HTML
                                                                       5

┌─────────────────┐ ┌──────┐   websocket connection opened
│ hello world   │ │ SEND │
└─────────────────┘ └──────┘

You sent hello world

The websockets package is a native HTML 5 Websocket implementation for the R language. The
package provides functions for setting up websocket servers and clients in R. It's
especially well-suited to lightweight interaction between R and web scripting languages
like Javascript. Multiple simultaneous websocket connections are supported.

The library has few external dependencies and is easily portable. More significantly,
websockets lets Javascript and other scripts embedded in web pages directly interact with
R, bypassing traditional middleware layers like .NET, Java, and web servers normally used
for such interaction. Although not its primary function, the package can serve basic HTTP
web pages like this one.

The HTML 5 Websocket API is a modern socket-like communication protocol for the web. It
can be much more efficient than Ajax and other polling methods. Note that the HTML 5
Websocket API is still under development and may change. Some browsers may not enable
Websockets by default and/or have quirky implementations (especially versions of Firefox),
but there are usually simple methods to enable the API. Despite its developmental status,
the API is presently widely supported: most recent browsers support it and there are many
available langauge implementations.

This simple example illustrates a few features of websockets. To see other available
demos, run the following in your R session:

demo(package='websockets')

websockets package Copyright (C) 2011 by Bryan W. Lewis, <blewis@illposed.net>, package licensed
under GNU LGPL v3.

Figure 5.4   Demo application of websockets.

5.4.4 Create a WebSocket Server Instance Using R

Next, we create a customized WebSocket server application.

```
# Open a new R program.
~ R

# Load class library.
> library(websockets)
```

```
# HTTP output of browser.
> text = "<html><body><h1>Hello world</h1></body></html>"

# Create service instance.
> w = create_server(port=7681,webpage=static_text_service(text))

# Listen "receive".
> recv = function(DATA, WS,...){
+  cat("Receive callback\n")
+  D = ""
+  if(is.raw(DATA)){D = rawToChar(DATA)}
+
+  cat("Callback:You sent",D,"\n")
+  websocket_write(DATA=paste("You sent",D,"\n",collapse=" "),WS=WS)
+}
> set_callback('receive',recv,w)

# Listen "closed".
> cl = function(WS){
+  cat("Websocket client socket ",WS$socket," has closed.\n")
+ }
> set_callback('closed',cl,w)

# Connection established
> es = function(WS){
+  cat("Websocket client socket ",WS$socket," has been
established.\n")
+}
> set_callback('established',es,w)

# Listen all connections
> while(TRUE) service(w)
```

Now we've create the server part of the WebSocket server.

5.4.5 Create a WebSocket Client Connection Using R

Then we create the server part of WebSocket application using R and communicate using the WebSocket server. First, create a new file client.r in Linux.

```
~ vi client.r

# Load class library.
library(websockets)

# Create client instance.
client = websocket("ws://192.168.1.201",port=7681)

# Listen "receive".
rece<-function(DATA, WS, HEADER)  {
```

```
    D=''
    if(is.raw(DATA)){
        cat("raw data")
      D = rawToChar(DATA)
    }
    cat("==>",D,"\n")
}
set_callback("receive",rece, client)

# Send request to server.
websocket_write("2222", client)

# Output the return value to server.
service(client)

# Close connection.
websocket_close(client)
```

Run the client program.

```
> library(websockets)
> client = websocket("ws://192.168.1.201",port=7681)
> rece<-function(DATA, WS, HEADER)  {
+    D=''
+    if(is.raw(DATA)){
+       cat("raw data")
+       D=rawToChar(DATA)
+    }
+    cat("==>",D,"\n")
+}
> set_callback("receive",rece, client)
> websocket_write("2222", client)
[1] 1

> service(client)
raw data ==> You sent 2222

> websocket_close(client)
Client socket 3 was closed.
```

It can be seen from the output that we've achieved the communication process between the client and the server.

5.4.6 Achieve Client Connection through HTML5 Native API of Browsers

Next, we write JavaScript code in the Chrome browser and call native HTML5 API to access WebSocket server. Open the page http://192.168.1.201:7681, press F12 to open developer tool, and switch to the Console of JavaScript. Write JavaScript code in Console, as in Figure 5.5.

Native HTML5 program:

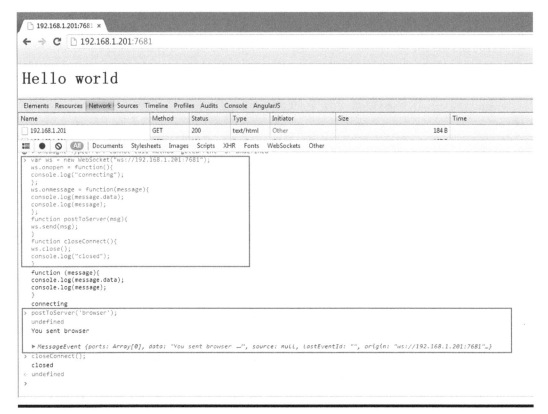

Figure 5.5   Use native JavaScript to access WebSocket server communication.

```
var ws = new WebSocket("ws://192.168.1.201:7681");
ws.onopen = function(){
console.log("connecting");
};
ws.onmessage = function(message){
console.log(message.data);
console.log(message);
};
function postToServer(msg){
ws.send(msg);
}
function closeConnect(){
ws.close();
console.log("closed");
}

postToServer('browser');
closeConnect();
```

Now we've finished the WebSocket server test constructed by R and opened a more convenient channel for R and other languages.

DATABASE AND BIG DATA

Chapter 6

Database and NoSQL

This chapter mainly introduces five tool packages of R accessing databases, which may help readers to connect R with MySQL, MongoDB, Redis, Cassandra, and Hive. The last section of this chapter introduces a case of financial big data based on Hive.

6.1 Programming Guidance for the RMySQL Database

Question

How do we use R to access MySQL?

MySQL is the most commonly used open source database software. It's simple to install (for the installation and configuration of MySQL, please refer to Appendix B) and stable in operation, which makes it very suitable for small and medium size data storage. R, as a data analysis tool, should certainly support a database driver interface. Great energy will be created if we combine R with MySQL.

6.1.1 Installation of RMySQL in Linux

Linux system environment:

- Linux: Ubuntu 12.04.2 LTS 64bit server
- Linux character set: en_US.UTF-8
- R: 3.0.1, x86_64-pc-linux-gnu (64-bit)
- MySQL: Ver 14.14 Distrib 5.5.29 64bit server
- MySQL character set: utf8

6.1.1.1 View the System Environment of Linux

```
# Linux kernel.
~ uname -a
Linux conan 3.5.0-23-generic #35~precise1-Ubuntu SMP Fri Jan 25
17:13:26 UTC 2013 x86_64 x86_64 x86_64 GNU/Linux

# Linux version.
~ cat /etc/issue
Ubuntu 12.04.2 LTS \n \l

# Linux system language.
~ locale
LANG=en_US.UTF-8
LANGUAGE=
LC_CTYPE="en_US.UTF-8"
LC_NUMERIC="en_US.UTF-8"
LC_TIME="en_US.UTF-8"
LC_COLLATE="en_US.UTF-8"
LC_MONETARY="en_US.UTF-8"
LC_MESSAGES="en_US.UTF-8"
LC_PAPER="en_US.UTF-8"
LC_NAME="en_US.UTF-8"
LC_ADDRESS="en_US.UTF-8"
LC_TELEPHONE="en_US.UTF-8"
LC_MEASUREMENT="en_US.UTF-8"
LC_IDENTIFICATION="en_US.UTF-8"
LC_ALL=en_US.UTF-8

# View the version of MySQL.
~ mysql --version
mysql  Ver 14.14 Distrib 5.5.29, for debian-linux-gnu (x86_64) using
readline 6.2

# View the character set of MySQL.
mysql> show variables like '%char%';
+--------------------------+----------------------------+
| Variable_name            | Value                      |
+--------------------------+----------------------------+
| character_set_client     | utf8                       |
| character_set_connection | utf8                       |
```

```
| character_set_database   | utf8                      |
| character_set_filesystem | binary                    |
| character_set_results    | utf8                      |
| character_set_server     | utf8                      |
| character_set_system     | utf8                      |
| character_sets_dir       | /usr/share/mysql/charsets/ |
+--------------------------+---------------------------+
8 rows in set (0.00 sec)
```

6.1.1.2 Install RMySQL in R

```
# Start R.
~ R

# Install RMySQL.
> install.packages('RMySQL')

# Omit parts of the output.

trying URL 'http://cran.dataguru.cn/src/contrib/DBI_0.2-7.tar.gz'
Content type 'application/x-gzip' length 194699 bytes (190 Kb)
opened URL
==================================================
downloaded 190 Kb

trying URL 'http://cran.dataguru.cn/src/contrib/RMySQL_0.9-3.tar.gz'
Content type 'application/x-gzip' length 165363 bytes (161 Kb)
opened URL
==================================================
downloaded 161 Kb


...


Configuration error:
  could not find the MySQL installation include and/or library
  directories. Manually specify the location of the MySQL
  libraries and the header files and re-run R CMD INSTALL.

INSTRUCTIONS:

1. Define and export the 2 shell variables PKG_CPPFLAGS and
   PKG_LIBS to include the directory for header files (*.h)
   and libraries, for example (using Bourne shell syntax):

      export PKG_CPPFLAGS="-I"
      export PKG_LIBS="-L -lmysqlclient"

   Re-run the R INSTALL command:

      R CMD INSTALL RMySQL_.tar.gz
```

```
2. Alternatively, you may pass the configure arguments
      --with-mysql-dir= (distribution directory)
   or
      --with-mysql-inc= (where MySQL header files reside)
      --with-mysql-lib= (where MySQL libraries reside)
   in the call to R INSTALL --configure-args='...'

   R CMD INSTALL --configure-args='--with-mysql-dir=DIR' RMySQL
_.tar.gz

ERROR: configuration failed for package 'RMySQL'
* removing '/home/conan/R/x86_64-pc-linux-gnu-library/3.0/RMySQL'

The downloaded source packages are in
        '/tmp/Rtmpu0Gn88/downloaded_packages'
Warning message:
In install.packages("RMySQL") :
  installation of package 'RMySQL' had non-zero exit status
```

The installation process reported errors and prompted that we should add the configuration parameter –with-mysql-dir to the installation directory of MySQL. Then let's solve the error.

```
# Install MySQL class library.
~ sudo apt-get install libdbd-mysql libmysqlclient-dev

# Find the installation directory of MySQL.
~ whereis mysql
mysql:/usr/bin/mysql/etc/mysql/usr/lib/mysql/usr/bin/X11/mysql/usr/
share/mysql/usr/share/man/man1/mysql.1.gz

# Find the downloaded file, RMySQL_.tar.gz.
~ ls/tmp/Rtmpu0Gn88/downloaded_packages
DBI_0.2-7.tar.gz RMySQL_0.9-3.tar.gz

# Re-install RMySQL through command line.
~ R CMD INSTALL --configure-args = '--with-mysql-dir=/usr/lib/
mysql'/tmp/Rtmpu0Gn88/downloaded_packages/RMySQL_0.9-3.tar.gz

* installing to library '/home/conan/R/
x86_64-pc-linux-gnu-library/3.0'
* installing *source* package 'RMySQL'...
** package 'RMySQL' successfully unpacked and MD5 sums checked

# Omit some of the output
installing to/home/conan/R/x86_64-pc-linux-gnu-library/3.0/RMySQL/
libs
** R
** inst
```

```
** preparing package for lazy loading
Creating a generic function for 'format' from package 'base' in
package 'RMySQL'
Creating a generic function for 'print' from package 'base' in
package 'RMySQL'
** help
*** installing help indices
** building package indices
** installing vignettes
** testing if installed package can be loaded
* DONE (RMySQL)
```

RMySQL is now successfully installed.

6.1.1.3 Create Library and Table in MySQL

```
# Log in local MySQL through MySQL command line client.

~ mysql -uroot -p

# Create a database, named rmysql.
mysql> create database rmysql;
Query OK, 1 row affected (0.00 sec)

# Create a user, username rmysql, password rmysql. Authorize this
user to operate RMySQL library remotely.
mysql> grant all on rmysql.* to rmysql@'%' identified by 'rmysql';
Query OK, 0 rows affected (0.00 sec)

# Authorize RMySQL user to operate RMySQL locally.
mysql> grant all on rmysql.* to rmysql@localhost identified by
'rmysql';
Query OK, 0 rows affected (0.00 sec)

# Switch to RMySQL library.
mysql> use rmysql
Database changed

# Create a table, t_user.
mysql> CREATE TABLE t_user(
    -> id INT PRIMARY KEY AUTO_INCREMENT,
    -> user varchar(12) NOT NULL UNIQUE
    -> )ENGINE=INNODB DEFAULT CHARSET=utf8;
Query OK, 0 rows affected (0.07 sec)

# Insert 3 items.
mysql> INSERT INTO t_user(user) values('A1'),('AB'),('fens.me');
Query OK, 3 rows affected (0.04 sec)
Records: 3  Duplicates: 0  Warnings: 0
```

```
# Check the data.
mysql> SELECT * FROM t_user;
+----+---------+
| id | user    |
+----+---------+
|  1 | A1      |
|  2 | AB      |
|  3 | fens.me |
+----+---------+
3 rows in set (0.00 sec)
```

6.1.1.4 Read MySQL Data through R

```
# Start R.
~ R

# Load RMySQL.
> library(RMySQL)
Loading required package: DBI

# Create Connection.
> conn <- dbConnect(MySQL(), dbname = "rmysql", username="rmysql",
password="rmysql")

# Run SQL.
> users = dbGetQuery(conn, "SELECT * FROM t_user")

# View data.
> users
  id     user
1  1       A1
2  2       AB
3  3 fens.me

# Disconnect.
> dbDisconnect(conn)
[1] TRUE
```

Now we've achieved the connection between R and MySQL in Linux Ubuntu.

6.1.2 Installation of RMySQL in Windows 7

Windows system environment:

- Windows 7 64 bit
- Windows character set: gbk, utf8
- R: 3.0.1, x86_64-w64-mingw32/x64 (64-bit)
- MySQL: mysql Ver 14.14 Distrib 5.6.11, for Win64 (x86_64)

6.1.2.1 View the System Environment of Windows 7

```
~ R --version
R version 3.0.1 (2013-05-16) -- "Good Sport"
Copyright (C) 2013 The R Foundation for Statistical Computing
Platform: x86_64-w64-mingw32/x64 (64-bit)

R is free software and comes with ABSOLUTELY NO WARRANTY.
You are welcome to redistribute it under the terms of the
GNU General Public License versions 2 or 3.
For more information about these matters see

http://www.gnu.org/licenses/.

# MySQL version.
~ mysql --version
mysql  Ver 14.14 Distrib 5.6.11, for Win64 (x86_64)

# MySQL character set.
mysql> show variables like '%char%';
+----------------------------+-------------------------------------+
| Variable_name              | Value                               |
+----------------------------+-------------------------------------+
| character_set_client       | gbk                                 |
| character_set_connection   | gbk                                 |
| character_set_database     | utf8                                |
| character_set_filesystem   | binary                              |
| character_set_results      | gbk                                 |
| character_set_server       | utf8                                |
| character_set_system       | utf8                                |
| character_sets_dir         | D:\toolkit\mysql56\share\charsets\  |
+----------------------------+-------------------------------------+
8 rows in set (0.07 sec)
```

6.1.2.2 Installation of RMySQL in R

```
# Start R.
~ R

# Install RMySQL.
> install.packages('RMySQl')
package 'RMySQl' is not available (for R version 3.0.1)
```

Here we have again met errors. It prompted that there was no corresponding version of RMySQL. Because RMySQL does not provide the distribution of Windows version, we have to compile the package of RMySQL manually in Windows and install it.

```
# Download the source code package of RMySQL.
> install.packages("RMySQL", type="source")
URLhttp://cran.dataguru.cn/src/contrib/RMySQL_0.9-3.tar.gz'
Content type 'application/x-gzip' length 165363 bytes (161 Kb)
URL
downloaded 161 Kb
```

Find the source code package RMySQL_0.9-3.tar.gz.

```
~ dir C:\Users\Administrator\AppData\Local\Temp\RtmpsfqQjK\
downloaded_packages
2013-09-24  13:16            165,363 RMySQL_0.9-3.tar.gz
```

Install it through source code.

```
~ D:\workspace\R\mysql>R CMD INSTALL C:\Users\Administrator\AppData\
Local\Temp\RtmpsfqQjK\downloaded_packages\RMySQL_0.9-3.tar.gz
* installing to library 'C:/Program Files/R/R-3.0.1/library'
* installing *source* package 'RMySQL' ...
** 'RMySQL'MD5
checking for $MYSQL_HOME... not found... searching registry...

# Omit parts of the output.
  MS-DOS style path detected: C:/PROGRA~1/R/R-30~1.1/bin/x64/Rscript
  Preferred POSIX equivalent is: /cygdrive/c/PROGRA~1/R/R-30~1.1/
bin/x64/Rscript
  CYGWIN environment variable option "nodosfilewarning" turns off
this warning.
  Consult the user's guide for more details about POSIX paths:

http://cygwin.com/cygwin-ug-net/using.html#using-pathnames

readRegistry("SOFTWARE\\MySQL AB", hive = "HLM", maxdepth = 2) :
  Registry key 'SOFTWARE\MySQL AB' not found

ERROR: configuration failed for package 'RMySQL'
* removing 'C:/Program Files/R/R-3.0.1/library/RMySQL'
```

There is an error again. We need to set the environment variables of MYSQL_HOME.

```
set MYSQL_HOME = D:\toolkit\mysql56
```

Note: we suggest that MYSQL_HOME be set in the system environment variables. Install RMySQL again.

```
D:\workspace\R\mysql>R CMD INSTALL C:\Users\Administrator\AppData\
Local\Temp\RtmpsfqQjK\downloaded_packages\RMySQL_0.9-3
.tar.gz
# Omit some output.
cc.exe: error: D:\toolkit\mysql56/bin/libmySQL.dll: No such file or
directory
ERROR: compilation failed for package 'RMySQL'
* removing 'C:/Program Files/R/R-3.0.1/library/RMySQL'
```

Then there is another error. This time it prompted that there was no dynamic linked library file D:\toolkit\mysql56\bin\libmySQL.dll.

```
# Copy dynamic linked library libmySQL.dll.
cp D:\toolkit\mysql56\lib\libmysql.dll D:\toolkit\mysql56\bin\
mv D:\toolkit\mysql56\bin\libmysql.dll D:\toolkit\mysql56\bin\
libmySQL.dll
```

Install RMySQL again.

```
~ D:\workspace\R\mysql>R CMD INSTALL C:\Users\Administrator\AppData\
Local\Temp\RtmpsfqQjK\downloaded_packages\RMySQL_0.9-3
.tar.gz
# Omit some output.
installing to C:/Program Files/R/R-3.0.1/library/RMySQL/libs/x64
** R
** inst
** preparing package for lazy loading
Creating a generic function for 'format' from package 'base' in
package 'RMySQL'
Creating a generic function for 'print' from package 'base' in
package 'RMySQL'
** help
*** installing help indices
** building package indices
** installing vignettes
** testing if installed package can be loaded
MYSQL_HOME defined as D:\toolkit\mysql56
* DONE (RMySQL)
```

The installation is finally successful!

6.1.2.3 Create Libraries and Tables in MySQL

```
~ mysql -uroot -p
mysql> create database rmysql;
Query OK, 1 row affected (0.04 sec)
```

```
mysql> grant all on rmysql.* to rmysql@'%' identified by 'rmysql';
Query OK, 0 rows affected (0.00 sec)

mysql> grant all on rmysql.* to rmysql@localhost identified by
'rmysql';
Query OK, 0 rows affected (0.00 sec)

mysql> use rmysql
Database changed
mysql> CREATE TABLE t_user(
    -> id INT PRIMARY KEY AUTO_INCREMENT,
    -> user varchar(12) NOT NULL UNIQUE
    -> )ENGINE=INNODB DEFAULT CHARSET=utf8;
Query OK, 0 rows affected (1.01 sec)

mysql> INSERT INTO t_user(user) values('A1'),('AB'),('fens.me');
Query OK, 3 rows affected (0.05 sec)
Records: 3  Duplicates: 0  Warnings: 0

mysql> SELECT * FROM t_user;
+----+---------+
| id | user    |
+----+---------+
|  1 | A1      |
|  2 | AB      |
|  3 | fens.me |
+----+---------+
3 rows in set (0.03 sec)
```

The process is in accordance with that of Linux.

```
~ mysql -uroot -p
mysql> create database rmysql;
Query OK, 1 row affected (0.04 sec)

mysql> grant all on rmysql.* to rmysql@'%' identified by 'rmysql';
Query OK, 0 rows affected (0.00 sec)

mysql> grant all on rmysql.* to rmysql@localhost identified by
'rmysql';
Query OK, 0 rows affected (0.00 sec)

mysql> use rmysql
Database changed
mysql> CREATE TABLE t_user(
    -> id INT PRIMARY KEY AUTO_INCREMENT,
    -> user varchar(12) NOT NULL UNIQUE
    -> )ENGINE=INNODB DEFAULT CHARSET=utf8;
Query OK, 0 rows affected (1.01 sec)

mysql> INSERT INTO t_user(user) values('A1'),('AB'),('fens.me');
```

```
Query OK, 3 rows affected (0.05 sec)
Records: 3  Duplicates: 0  Warnings: 0

mysql> SELECT * FROM t_user;
+----+---------+
| id | user    |
+----+---------+
|  1 | A1      |
|  2 | AB      |
|  3 | fens.me |
+----+---------+
3 rows in set (0.03 sec)
```

6.1.2.4 Read MySQL Data through R

If you haven't written the MYSQL_HOME variables in the environment configuration, you need to set the variable every time you start R.

```
# Set the environment variable in the current environment.
~ set MYSQL_HOME=D:\toolkit\mysql56

# Start R.
~ R

# Load RMySQL.
> library(RMySQL)
DBI
MYSQL_HOME defined as D:\toolkit\mysql56

# Create connection.
> conn <- dbConnect(MySQL(), dbname = "rmysql", username="rmysql",
password="rmysql")

# Run SQL.
> users = dbGetQuery(conn, "SELECT * FROM t_user")
> users
  id    user
1  1      A1
2  2      AB
3  3 fens.me

# Disconnect.
> dbDisconnect(conn)
[1] TRUE
```

We have now achieved the connection between R and MySQL in Windows.

6.1.3 Use of RMySQL Functions

After the environment is installed, let's begin to use the RMySQL package. This section mainly introduces the auxiliary operation, database operation, and character set configuration of RMy SQL in Windows.

6.1.3.1 Auxiliary Function of RMySQL

```
# Create local connection.
> conn <- dbConnect(MySQL(), dbname = "rmysql", username="rmysql",
password="rmysql",client.flag=CLIENT_MULTI_STATEMENTS)

# Create remote connection.
> conn <- dbConnect(MySQL(), dbname = "rmysql", username="rmysql",
password="rmysql",host="192.168.1.201",port=3306)

# Close connection.
> dbDisconnect(conn)

# View all the tables in the database.
> dbListTables(conn)
[1] "t_user"

# View the field of table.
> dbListFields(conn, "t_user")
[1] "id"    "user"

# Check the MySQL information.
> summary(MySQL(), verbose = TRUE)
<MySQLDriver:(23864)>
  Driver name:  MySQL
  Max  connections: 16
  Conn. processed: 3
  Default records per fetch: 500
  DBI API version:

# Information of MySQL connection instance.
> summary(conn, verbose = TRUE)
<MySQLConnection:(23864,2)>
  User: root
  Host: localhost
  Dbname: rmysql
  Connection type: localhost via TCP/IP
  MySQL server version:  5.6.11
  MySQL client version:  5.6.11
  MySQL protocol version:  10
  MySQL server thread id:  35
  No resultSet available

# Information of MySQL connection.
> dbListConnections(MySQL())
[[1]]
<MySQLConnection:(23864,2)>
```

6.1.3.2 Operation of RMySQL Database

```
# Create table and insert data.
> t_demo<-data.frame(
  a=seq(1:10),
  b=letters[1:10],
  c=rnorm(10)
)
> dbWriteTable(conn, "t_demo", t_demo)

# Acquire the data of the whole table.
> dbReadTable(conn, "t_demo")
    a b          c
1   1 a  0.98868164
2   2 b -0.66935770
3   3 c  0.27703638
4   4 d  1.36137156
5   5 e -0.70291017
6   6 f  1.61235088
7   7 g  0.17616068
8   8 h  0.29700017
9   9 i  0.19032719
10 10 j -0.06222173

# Insert new data.
> dbWriteTable(conn, "t_demo", t_demo, append=TRUE)
> dbReadTable(conn, "t_demo")
   row_names  a b          c
1          1  1 a  0.98868164
2          2  2 b -0.66935770
3          3  3 c  0.27703638
4          4  4 d  1.36137156
5          5  5 e -0.70291017
6          6  6 f  1.61235088
7          7  7 g  0.17616068
8          8  8 h  0.29700017
9          9  9 i  0.19032719
10        10 10 j -0.06222173
11         1  1 a  0.98868164
12         2  2 b -0.66935770
13         3  3 c  0.27703638
14         4  4 d  1.36137156
15         5  5 e -0.70291017
16         6  6 f  1.61235088
17         7  7 g  0.17616068
18         8  8 h  0.29700017
19         9  9 i  0.19032719
20        10 10 j -0.06222173

# Cover the data of the old table.
> dbWriteTable(conn, "t_demo", t_demo, overwrite=TRUE)

# Query of data.
```

```
> d0 <- dbGetQuery(conn, "SELECT * FROM t_demo where c>0")
> class(d0)
[1] "data.frame"

> d0
  row_names a b         c
1         1 1 a 0.9886816
2         3 3 c 0.2770364
3         4 4 d 1.3613716
4         6 6 f 1.6123509
5         7 7 g 0.1761607
6         8 8 h 0.2970002
7         9 9 i 0.1903272

# Run SQL script query and make paging.
> rs <- dbSendQuery(conn, "SELECT * FROM t_demo where c>0")
> class(rs)
[1] "MySQLResult"
attr(,"package")
[1] "RMySQL"
> mysqlCloseResult(rs)
[1] TRUE

> d1 <- fetch(rs, n = 3)
> d1
  row_names a b         c
1         1 1 a 0.9886816
2         3 3 c 0.2770364
3         4 4 d 1.3613716

# View the statistical information of the set.
> summary(rs, verbose = TRUE)
  row_names                a               b               c
 Length:7        Min.   :1.000    Length:7        Min.   :0.1762
 Class :character 1st Qu.:3.500    Class :character 1st Qu.:0.2337
 Mode  :character Median :6.000    Mode  :character Median :0.2970
                  Mean   :5.429                    Mean   :0.7004
                  3rd Qu.:7.500                    3rd Qu.:1.1750
                  Max.   :9.000                    Max.   :1.6124

# Not insert row.names field.
> dbWriteTable(conn, "t_demo", t_demo,row.names=FALSE,overwrite=TRUE)
> dbGetQuery(conn, "SELECT * FROM t_demo where c>0")
  a b         c
1 1 a 0.9886816
2 3 c 0.2770364
3 4 d 1.3613716
4 6 f 1.6123509
5 7 g 0.1761607
```

```
6  8 h 0.2970002
7  9 i 0.1903272

# Delete table.
> if(dbExistsTable(conn,'t_demo')){
+       dbRemoveTable(conn, "t_demo")
+ }
[1] TRUE
```

    Run SQL statement, dbSendQuery.

```
> query<-dbSendQuery(conn, "show tables")
> data <- fetch(query, n = -1)
> data
  Tables_in_rmysql
1          t_demo
2          t_user
> mysqlCloseResult(query)
[1] TRUE
```

Special tip: Try not to use dbWriteTable(), as it will delete the original table structure, create a new table structure, and insert data.

6.1.3.3 Character Set Configuration of Windows

When we insert Chinese characters in MySQL in Windows 7, there will be garbled words as the character encoding is GBK. Here is the way to insert Chinese characters through the MySQL command line client.

```
# Insert data of Chinese characters.
mysql> INSERT INTO t_user(user) values('小朋友'),('你好'),('正确了');
Query OK, 3 rows affected (0.07 sec)
Records: 3  Duplicates: 0  Warnings: 0

# The Chinese characters are displayed correctly.
mysql> select * from t_user;
+----+---------+
| id | user    |
+----+---------+
|  1 | A1      |
|  2 | AB      |
|  3 | fens.me |
|  5 | 你好    |
|  4 | 小朋友  |
|  6 | 正确了  |
+----+---------+
6 rows in set (0.07 sec)
```

Query the data through RMySQL.

```
# The Chinese characters are garbled.
> dbGetQuery(conn, "SELECT * FROM t_user")
  id     user
1  1       A1
2  2       AB
3  3 fens.me
4  5       ??
5  4      ???
6  6      ???
```

Then, set the GBK character set in the API of R.

```
> dbDisconnect(conn)
> conn <- dbConnect(MySQL(), dbname = "rmysql", username="root",
password="",client.flag=CLIENT_MULTI_STATEMENTS)

# Set the GBK character set.
> dbSendQuery(conn,'SET NAMES gbk')

# Query the data.
> query<-dbSendQuery(conn, "SELECT * FROM t_user")
> data <- fetch(query, n = -1)
> mysqlCloseResult(query)
[1] TRUE

# Chinese characters are displayed correctly now.
> data
  id     user
1  1       A1
2  2       AB
3  3 fens.me
4  5      你好
5  4    小朋友
6  6    正确了
```

6.1.4 Case Practice of RMySQL

Next, let's practice a case of a remote database connection. Use RMySQL client in Windows to connect the MySQL database server in Linux remotely.

6.1.4.1 Create Tables in a Remote Database

Create a new table of primary key index or unique key index, t_blog, in MySQL using an SQL statement. Here is the statement.

```
# Create table.
mysql> CREATE TABLE t_blog(
id INT PRIMARY KEY AUTO_INCREMENT,
title varchar(12) NOT NULL UNIQUE,
author varchar(12) NOT NULL,
length int NOT NULL,
create_date timestamp NOT NULL DEFAULT now()
)ENGINE=INNODB DEFAULT CHARSET=UTF8;


# View the table structure.
mysql> desc t_blog;
+-------------+-------------+------+-----+-------------------+----------------+
| Field       | Type        | Null | Key | Default           | Extra          |
+-------------+-------------+------+-----+-------------------+----------------+
| id          | int(11)     | NO   | PRI | NULL              | auto_increment |
| title       | varchar(12) | NO   | UNI | NULL              |                |
| author      | varchar(12) | NO   |     | NULL              |                |
| length      | int(11)     | NO   |     | NULL              |                |
| create_date | timestamp   | NO   |     | CURRENT_TIMESTAMP |                |
+-------------+-------------+------+-----+-------------------+----------------+
5 rows in set (0.00 sec)


# View table index.
mysql> show indexes from t_blog;
+--------+------------+----------+--------------+-------------+-----------+-------
------+----------+--------+------+------------+---------+---------------+---------
| Table  | Non_unique | Key_name | Seq_in_index | Column_name | Collation
| Cardinality | Sub_part | Packed | Null | Index_type | Comment | Index_comment |
+--------+------------+----------+--------------+-------------+-----------+--
-----------+----------+--------+------+------------+---------+---------------+
| t_blog | 0 | PRIMARY  | 1 |   id  | A | 3 | NULL | NULL |   | BTREE |
|   |
| t_blog | 0 | title    | 1 | title | A | 3 | NULL | NULL |   | BTREE |
|   |
+--------+------------+----------+--------------+-------------+-----------+-------
------+----------+--------+------+------------+---------+---------------+---------
2 rows in set (0.00 sec)


# Insert data.
mysql> INSERT INTO t_blog(title,author,length) values('Hello, this is the
first article','Conan',20),('Coding of RMySQL','Conan',99),('R for
Programmers series','Conan',15);


# Query table.
mysql> select * from t_blog;
+----+-------------------------------+--------+--------+---------------------+
| id | title                         | author | length | create_date         |
+----+-------------------------------+--------+--------+---------------------+
|  1 | Hello, this is the first article | Conan | 20 | 2013-08-15 00:13:13 |
|  2 | Coding of RMySQL              | Conan  |     99 | 2013-08-15 00:13:13 |
|  3 | R for Programmers series      | Conan  |     15 | 2013-08-15 00:13:13 |
+----+-------------------------------+--------+--------+---------------------+
3 rows in set (0.00 sec)
```

6.1.4.2 Use RMySQL to Access MySQL Remotely

Use RMySQL to access the database remotely. Insert data and take data.

```
> library(RMySQL)

# Create remote connection.
> conn <- dbConnect(MySQL(), dbname = "rmysql",
username="rmysql", password="rmysql",host="192.168.1.201",port=3306)

# Set the GBK character encoding.
> dbSendQuery(conn,'SET NAMES gbk')

# Run SQL.
> dbSendQuery(conn,"INSERT INTO t_blog(title,author,length) values
('Insert new article in R','Conan',50)");

# Query data.
> query<-dbSendQuery(conn, "SELECT * FROM t_blog")
Warning message:
In mysqlExecStatement(conn, statement, ...) :
  RS-DBI driver warning: (unrecognized MySQL field type 7 in column 4 imported
as character)

# Query data.
> data <- fetch(query, n = -1)
> mysqlCloseResult(query)
[1] TRUE

# Display is normal.
> print(data)
  id              title author length      create_date
1  1   Hello, this is the first article  Conan   20 2013-08-15 00:13:13
2  2   Coding of RMySQL                  Conan   99 2013-08-15 00:13:13
3  3   R for Programmers series          Conan   15 2013-08-15 00:13:13
4  4   Insert new article in R           Conan   50 2013-08-15 00:29:45

> dbDisconnect(conn)
[1] TRUE
```

We can reduce the error and increase working efficiency by mastering all the techniques of RMySQL and understanding the principle.

6.2 Connecting R with NoSQL: MongoDB

Question

How do we use R to connect MongoDB?

MongoDB, as a documental NoSQL database, is very flexible in use. It avoids the complex database design of relational database in the early stage. The storage of MongoDB is based on JSON, and adopts JavaScript as an operating language, which gives the user infinite space for imagination. We can solve some really complex questions of criteria query by coding in MongoDB server. This section will introduce how to connect R with MongoDB through rmongodb.

6.2.1 Environment Preparation of MongoDB

For environment preparation, I choose Linux® Ubuntu here. You may choose other types of Linux according to your preference.

- Linux: Ubuntu 12.04.2 LTS 64bit server
- MongoDB: v2.4.9
- IP: 192.168.1.199

About the installation and configuration of MongoDB, please refer to Appendix D. Then let's view the server environment of MongoDB. Use command/etc/init.d/mongodb to start Mongo DB. The default port is 27017.

```
# Start mongodb.
~ sudo /etc/init.d/mongodb start
Rather than invoking init scripts through /etc/init.d, use the service(8)
utility, e.g. service mongodb start

Since the script you are attempting to invoke has been converted to an
Upstart job, you may also use the start(8) utility, e.g. start mongodb
mongodb start/running, process 1878

# View the system process.
~ ps -aux|grep mongo
mongodb    1878  3.3  0.4 348168 37488 ?         Ssl  10:19   0:01 /usr/bin/
mongod --config /etc/mongodb.conf

# View the starting log.
~ cat /var/log/mongodb/mongodb.log
```

```
Mon Mar 31 10:19:28.764 [initandlisten] MongoDB starting : pid=1878
port=27017 dbpath=/var/lib/mongodb 64-bit host=rbook
Mon Mar 31 10:19:28.764 [initandlisten] db version v2.4.9
Mon Mar 31 10:19:28.764 [initandlisten] git version:
52fe0d21959e32a5bdbecdc62057db386e4e029c
Mon Mar 31 10:19:28.764 [initandlisten] build info: Linux ip-10-2-29-40
2.6.21.7-2.ec2.v1.2.fc8xen #1 SMP Fri Nov 20 17:48:28 EST 2009 x86_64
BOOST_LIB_VERSION=1_49
Mon Mar 31 10:19:28.764 [initandlisten] allocator: tcmalloc
Mon Mar 31 10:19:28.764 [initandlisten] options: { config: "/etc/mongodb.
conf", dbpath: "/var/lib/mongodb", logappend: "true", logpath: "/var/log/
mongodb/mongodb.log" }
Mon Mar 31 10:19:28.853 [initandlisten] journal dir=/var/lib/mongodb/journal
Mon Mar 31 10:19:28.853 [initandlisten] recover : no journal files present,
no recovery needed
Mon Mar 31 10:19:28.872 [initandlisten] waiting for connections on port 27017
Mon Mar 31 10:19:28.872 [websvr] admin web console waiting for connections on
port 28017
```

Configuration of MongoDB at runtime:

- Process number: pid = 1878
- Port: port = 27017
- Data file directory: dbpath =/var/lib/mongodb
- Host name: host = rbook
- IP: 192.168.1.199

Use the command line client program of MongoDB, mongo, to open Mongo Shell. Some simple operations of Mongo Shell include viewing the database, switching the database, and viewing the data set.

```
# Open mongo shell.
~ mongo
MongoDB shell version: 2.0.6
connecting to: test

# Enter mongo shell. Database displayed in a list.
> show dbs
db       0.0625GB
feed     0.0625GB
foobar   0.0625GB
local    (empty)

# Switch database.
> use foobar
switched to db foobar

# Data set displayed in a list.
> show collections
blog
system.indexes
```

Then, we'll use MongoDB client of R, rmongodb, to access the MongoDB server remotely to make a test.

6.2.2 Rmongodb Function Library

Rmongodb is a client communication interface program for R to access MongoDB. Rmongodb provides 153 functions, corresponding to different operations of MongoDB. Compared to other NoSQL packages, this project is really large. Though the amount is large, the use of these functions is rather simple. They are flexible in supporting R and concise in codes. I'll not list all of the 153 functions, but only select some commonly used functions to introduce. Readers who are interested in them can find all the functions in the official documents of rmongodb.

```
# Create mongo connection.
mongo<-mongo.create()
# Check whether the connection is normal.
mongo.is.connected(mongo)
# Create a BSON object chache.
buf <- mongo.bson.buffer.create()
# Add element to object buf.
mongo.bson.buffer.append(buf, "name", "Echo")
```

Add element to object class.

```
score <- c(5, 3.5, 4)
names(score) <- c("Mike", "Jimmy", "Ann")
mongo.bson.buffer.append(buf, "score", score)
```

Add element to array class.

```
mongo.bson.buffer.start.array(buf, "comments")
mongo.bson.buffer.append(buf, "0", "a1")
mongo.bson.buffer.append(buf, "1", "a2")
mongo.bson.buffer.append(buf, "2", "a3")
```

Close element of array class.

```
mongo.bson.buffer.finish.object(buf)
```

Take out cache data.

```
b <- mongo.bson.from.buffer(buf)
```

Database and data set.

```
ns = "db.blog"
```

Insert a record.

```
mongo.insert(mongo,ns,b)

#mongo shell:(Not Run)
db.blog.insert(b)
```

Create query object.

```
buf <- mongo.bson.buffer.create()
mongo.bson.buffer.append(buf, "name", "Echo")
query <- mongo.bson.from.buffer(buf)
```

Create query return value object.

```
buf <- mongo.bson.buffer.create()
mongo.bson.buffer.append(buf, "name", 1)
fields <- mongo.bson.from.buffer(buf)
```

Perform query of single record.

```
mongo.find.one(mongo, ns, query, fields)
#mongo shell:(Not Run)
db.blog.findOne({query},{fields})
```

Perform query of list record.

```
mongo.find(mongo, ns, query, fields)

#mongo shell:(Not Run)
db.blog.find({query},{fields})
```

Create the modifier object objNew.

```
buf <- mongo.bson.buffer.create()
mongo.bson.buffer.start.object(buf, "$inc")
mongo.bson.buffer.append(buf, "age", 1L)
mongo.bson.buffer.finish.object(buf)
objNew <- mongo.bson.from.buffer(buf)
```

Perform the modification operation.

```
mongo.update(mongo, ns, query, objNew)

#mongo shell:(Not Run)
db.blog.update({query},{objNew})
```

Modification of the single line of code.

```
mongo.update(mongo, ns, query, list(name="Echo", age=25))

#mongo shell:(Not Run)
db.blog.update({query},{objNew})
```

Delete the selected object.

```
mongo.remove(mongo, ns, query)

#mongo shell:(Not Run)
db.blog.remove({query},{objNew})
```

Destroy the mongo connection.

```
mongo.destroy(mongo)
```

6.2.3 Basic Operation of Rmongodb

Client environment of R:

- Win7 64bit
- R: 3.0.1 x86_64-w64-mingw32/x64 b4bit

Section 6.2.2 introduced some basic functions of rmongodb function library, but we have not installed the class library of rmongodb.

```
# Start R.
~ R

# Install rmongodb.
> install.packages("rmongodb")

# Load class library.
> library(rmongodb)
```

Then, use mongo.create() to create a connection with MongoDB server. If it's a local connection, mongo.create() won't need parameters. The next example uses remote connection. It'll add host parameters to specify the IP address of the MongoDB server.

```
# Connect mongodb server remotely.
> mongo<-mongo.create(host="192.168.1.199")
```

Use mongo.is.connected() to check whether the connection is normal. This statement will be used a great deal in development. If the object or function reports an error in R modeling, the connection will be closed automatically. It will not prompt disconnection because of the abnormal mechanism of MongoDB. We need to use this command to test whether the connection is normal manually.

```
# Check whether the connection is normal.
> print(mongo.is.connected(mongo))
[1] TRUE
```

Next, define two variables, db and ns. db is the database we use, and ns is the database plus data set.

```
# Define db.
> db<-"foobar"

# Define db.collection.
> ns<-"foobar.blog"
```

Next, we create a JSON object and save it to MongoDB.

```
{
    "_id" : ObjectId("51663e14da2c51b1e8bc62eb"),
    "name" : "Echo",
    "age" : 22,
    "gender" : "Male",
    "score" : {
            "Mike" : 5,
            "Jimmy" : 3.5,
            "Ann" : 4
    },
```

```
    "comments" : [
            "a1",
            "a2",
            "a3"
    ]
}
```

Organize class BSON data.

```
> buf <- mongo.bson.buffer.create()
> mongo.bson.buffer.append(buf, "name", "Echo")
> mongo.bson.buffer.append(buf, "age", 22L)
> mongo.bson.buffer.append(buf, "gender", 'Male')
```

Object class.

```
> score <- c(5, 3.5, 4)
> names(score) <- c("Mike", "Jimmy", "Ann")
> mongo.bson.buffer.append(buf, "score", score)
```

Array class.

```
> mongo.bson.buffer.start.array(buf, "comments")
> mongo.bson.buffer.append(buf, "0", "a1")
> mongo.bson.buffer.append(buf, "1", "a2")
> mongo.bson.buffer.append(buf, "2", "a3")
> mongo.bson.buffer.finish.object(buf)
> b <- mongo.bson.from.buffer(buf)  # Insert to mongodb.
> mongo.insert(mongo,ns,b)
```

Display single inserted data.

```
> buf <- mongo.bson.buffer.create()
> mongo.bson.buffer.append(buf, "name", "Echo")
> query <- mongo.bson.from.buffer(buf)
> print(mongo.find.one(mongo, ns, query))
```

Then, use the modifiers $inc, $set, and $push to operate. First use $inc to add 1 to age.

```
> buf <- mongo.bson.buffer.create()
> mongo.bson.buffer.start.object(buf, "$inc")
> mongo.bson.buffer.append(buf, "age", 1L)
> mongo.bson.buffer.finish.object(buf)
> objNew <- mongo.bson.from.buffer(buf)
> mongo.update(mongo, ns, query, objNew)
> print(mongo.find.one(mongo, ns, query))
```

Use $set to set age = 1.

```
> buf <- mongo.bson.buffer.create()
> mongo.bson.buffer.start.object(buf, "$set")
> mongo.bson.buffer.append(buf, "age", 1L)
> mongo.bson.buffer.finish.object(buf)
> objNew <- mongo.bson.from.buffer(buf)
> mongo.update(mongo, ns, query, objNew)
> print(mongo.find.one(mongo, ns, query))
```

Use $push to add "Orange" data to the comments array.

```
> buf <- mongo.bson.buffer.create()
> mongo.bson.buffer.start.object(buf, "$push")
> mongo.bson.buffer.append(buf, "comments", "Orange")
> mongo.bson.buffer.finish.object(buf)
> objNew <- mongo.bson.from.buffer(buf)
> mongo.update(mongo, ns, query, objNew)
> print(mongo.find.one(mongo, ns, query))
```

Use simplified statements to reassign value to the object.

```
> mongo.update(mongo, ns, query, list(name="Echo", age=25))
> print(mongo.find.one(mongo, ns, query))
```

Last, delete object and disconnect.

```
# Delete object.
> mongo.remove(mongo, ns, query)
# Destroy mongo connection.
> mongo.destroy(mongo)
```

6.2.4 A Case of Performance Test of Rmongodb

Now that we've understood the use of rmongodb, let's run a case of performance test: insert data in batch, use a modifier to modify data in batch, and compare the speed of the three modifiers. First, insert data functions in batch.

```
# Load stringr to achieve character string operation.
> library(stringr)
# Insert data in batch.
> batch_insert<-function(arr=1:10,ns){
```

```
+     mongo_insert<-function(x){
+       buf <- mongo.bson.buffer.create()
+       mongo.bson.buffer.append(buf, "name", str_c("Dave",x))
+       mongo.bson.buffer.append(buf, "age", x)
+       mongo.bson.buffer.start.array(buf, "comments")
+       mongo.bson.buffer.append(buf, "0", "a1")
+       mongo.bson.buffer.append(buf, "1", "a2")
+       mongo.bson.buffer.append(buf, "2", "a3")
+       mongo.bson.buffer.finish.object(buf)
+       return(mongo.bson.from.buffer(buf))
+     }
+     mongo.insert.batch(mongo, ns, lapply(arr,mongo_insert))
+}
```

Modification in batch of modifier function: $inc.

```
> batch_inc<-function(data,ns){
+     for(i in data){
+       buf <- mongo.bson.buffer.create()
+       mongo.bson.buffer.append(buf, "name", str_c("Dave",i))
+       criteria <- mongo.bson.from.buffer(buf)
+       buf <- mongo.bson.buffer.create()
+       mongo.bson.buffer.start.object(buf, "$inc")
+       mongo.bson.buffer.append(buf, "age", 1L)
+       mongo.bson.buffer.finish.object(buf)
+       objNew <- mongo.bson.from.buffer(buf)
+       mongo.update(mongo, ns, criteria, objNew)
+     }
+}
```

Modification in batch of modifier function: $set.

```
> batch_set<-function(data,ns){
+     for(i in data){
+       buf <- mongo.bson.buffer.create()
+       mongo.bson.buffer.append(buf, "name", str_c("Dave",i))
+       criteria <- mongo.bson.from.buffer(buf)
+       buf <- mongo.bson.buffer.create()
+       mongo.bson.buffer.start.object(buf, "$set")
+       mongo.bson.buffer.append(buf, "age", 1L)
+       mongo.bson.buffer.finish.object(buf)
+       objNew <- mongo.bson.from.buffer(buf)
+       mongo.update(mongo, ns, criteria, objNew)
+     }
+ }
```

Modification in batch of modifier function: $push.

```
> batch_push<-function(data,ns){
+    for(i in data){
+       buf <- mongo.bson.buffer.create()
+       mongo.bson.buffer.append(buf, "name", str_c("Dave",i))
+       criteria <- mongo.bson.from.buffer(buf)
+       buf <- mongo.bson.buffer.create()
+       mongo.bson.buffer.start.object(buf, "$push")
+       mongo.bson.buffer.append(buf, "comments", "Orange")
+       mongo.bson.buffer.finish.object(buf)
+       objNew <- mongo.bson.from.buffer(buf)
+       mongo.update(mongo, ns, criteria, objNew)
+    }
+}
```

All of the three aforementioned modifier programs use for loop in statements, so their performance loss in for loop should be the same. Thus we'll not take this factor into consideration. Now let's write the execution program and compare the speed of these three modifiers.

```
# Assign the table.
> ns="foobar.blog"

# Time of loop.
> data=1:1000

# Clear data.
> mongo.remove(mongo, ns)
[1] TRUE

# Insert in batch.
> system.time(batch_insert(data, ns))
user   system elapsed
0.25    0.00    0.28

# Modifier $inc.
> system.time(batch_inc(data, ns))
user   system elapsed
0.47    0.27    2.50

# Modifier $se.
> system.time(batch_set(data, ns))
user   system elapsed
0.77    0.48    3.17

# Modifier $push.
> system.time(batch_push(data, ns))
user   system elapsed
0.81    0.41    4.23
```

The speed of these three modifiers is: $push > $set > $inc. Because each of these three modifiers operates on a different class, $push on array, $set on any value and $inc on numbers, all the test results will only serve as a reference.

We introduced how to use R to access JSON data in Section 1.6, and now we've successfully connected R with MongoDB. As the data storage type of MongoDB is JSON, the semistructured data processing basis of JSON is completed, and R will also play an important role in the area of semistructured data processing.

6.3 Connecting R with NoSQL: Redis

Question

How do we use R to connect Redis?

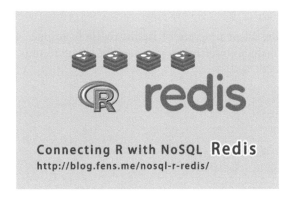

Redis is a Key-Value database based on RAM. It's more advanced than Memcache as it supports many data structures and it's efficient and fast. Redis can easily solve the problem of high concurrency data access. And it also performs well in real-time data storage. This section introduces how to connect Redis using R.

6.3.1 Environment Preparation of Redis

For environment preparation, I choose Linux Ubuntu here. You may choose other types of Linux according to your preference.

- Linux: Ubuntu 12.04.2 LTS 64bit server
- MongoDB: 2.2.12
- IP: 192.168.1.101

For information on the installation and configuration of Redis, please refer to Appendix C. Use command/etc/init.d/redis-server to start redis-server to view the server environment of Redis. The default port is 6379.

```
# Start redis.
~ /etc/init.d/redis-server start
Starting redis-server: redis-server.

# View system process.
~ ps -aux|grep redis
```

```
redis     20128  0.0  0.0  10676  1428 ?         Ss   16:39   0:00 /
usr/bin/redis-server /etc/redis/redis.conf

# View starting log.
~ cat  /var/log/redis/redis-server.log

[20128] 14 Apr 16:39:43 * Server started, Redis version 2.2.12
[20128] 14 Apr 16:39:43 # WARNING overcommit_memory is set to 0!
Background save may fail under low memory condition. To fix this
issue add 'vm.overcommit_memory = 1' to /etc/sysctl.conf and then
reboot or run the command 'sysctl vm.overcommit_memory=1' for this
to take effect.
```

Use the command line client program of Redis, redis-cli, to open Redis Shell. The simple operation of Redis Shell includes inserting a record and querying records.

```
# Open Redis Shell.
~ redis-cli

# View the use of set command.
redis 127.0.0.1:6379> help set

  SET key value
  summary: Set the string value of a key
  since: 0.07
  group: string

# Insert a record, key = a, value = 178
redis 127.0.0.1:6379> set a 178
OK
# View the record.
redis 127.0.0.1:6379> get a
"178"
```

Then, let's use the Redis cliet of R, rredis, to access the Redis server remotely to run the test.

6.3.2 Function Library of rredis

rredis is a client communication interface program for R to access Redis. It provides 100 functions, corresponding to different operations of Redis. Though the amount is rather large, the use of these functions is relatively simple. They are rather flexible in supporting R and concise in codes. I won't list all the 100 rredis functions but only select some frequently used functions to introduce. If you are interested in those functions, you may find them in the official documents of rredis.

```
# Create connection, close connection.
redisConnect() , redisClose()
# Flush the current/all data of database.
redisFlushDB() , redisFlushAll()
# List all the key value and number of keys.
redisKeys(), redisDBSize()
# Select database: 0 is the default database.
redisSelect(0)
#Insert string object, insert in batch.
redisSet('x',runif(5)), redisMSet(list(x=pi,y=runif(5),z=sqrt(2)))
# Read string object, read inbatch.
redisGet('x'), redisMGet(c('x','y','z'))
# Delete object.
redisDelete('x')
# Insert array object in the left, insert array object in the right.
redisLPush('a',1), redisRPush('a','A')
# Pop an array object in the left, pop an array object in the right.
redisLPop('a'), redisRPop('a')
#Display the array objects list from left.
redisLRange('a',0,-1)
# Insert object of class set.
redisSAdd('A',runif(2))
# Display the number of elements of set object, display the set
object elements in list.
redisSCard('A'), redisSMembers('A')
# Display the different set, intersection and union of two set
objects.
redisSDiff(c('A','B')),redisSInter(c('A','B')),redisSUnion(c('A',
'B'))
```

6.3.3 Basic Operation of rredis

Client environment of R:

- Win7 64bit
- R: 3.0.1 x86_64-w64-mingw32/x64 b4bit

Now I'll provide an introduction according to the following five operation types.

- Basic operation of redis: create connection, switch database, display all KEY value in list, clear the data of current database, close connection
- Operation of class string: insert, read, delete, insert and set expiration time, operation in batch
- Operation of class list: insert, read, pop
- Operation of class set: insert, read, different set, intersection, union
- Interactive operation between rredis and redis-cli

6.3.3.1 Basic Operation of Redis

```
# Install rredis.
> install.packages("rredis")

# Load rredis class library.
> library(rredis)
```

Create a connection with the Redis server through redisConnect(). If it's a local connection, redisConnect() won't need parameters. The following example uses a remote connection. Add the host parameter and configure the IP address redisConnect(host = "192.168.1.101",port = 6379).

```
# Connect redis server remotely.
> redisConnect(host="192.168.1.101",port=6379)

# List all the keys.
> redisKeys()
[1] "x"     "data"

# Display the number of keys.
> redisDBSize()
[1] 2

# Switch to database1.
> redisSelect(1)
[1] "OK"
> redisKeys()
NULL

# Switch to database0.
> redisSelect(0)
[1] "OK"
> redisKeys()
[1] "x"     "data"

# Clear the data of current database.
> redisFlushDB()
[1] "OK"

# Clear the data of all databases.
> redisFlushAll()
[1] "OK"

# Close connection.
> redisClose()
```

6.3.3.2 Operation of Class String

```
# Insert object.
> redisSet('x',runif(5))
[1] "OK"

# Read object.
> redisGet('x')
[1] 0.67616159 0.06358643 0.07478021 0.32129140 0.16264615

# Set the expiration time of data.
> redisExpire('x',1)
Sys.sleep(1)
> redisGet('x')
NULL

# Insert in batch.
> redisMSet(list(x=pi,y=runif(5),z=sqrt(2)))
[1] TRUE

# Read in batch.
> redisMGet(c('x','y','z'))
$x
[1] 3.141593
$y
[1] 0.9249501 0.3444994 0.6477250 0.1681421 0.2646853
$z
[1] 1.414214

# Delete data.
> redisDelete('x')
[1] 1
> redisGet('x')
NULL
```

6.3.3.3 Operation of Class List

```
# Insert data from the left of array.
> redisLPush('a',1)
> redisLPush('a',2)
> redisLPush('a',3)

# Display data of 0~2 from the left of array.
> redisLRange('a',0,2)
[[1]]
[1] 3
[[2]]
[1] 2
[[3]]
[1] 1
```

```
# Pop a data from the left of data.
> redisLPop('a')
[1] 3

# Display the data of 0~(-1) from the left of array.
> redisLRange('a',0,-1)
[[1]]
[1] 2
[[2]]
[1] 1

# Insert data from the right of array.
> redisRPush('a','A')
> redisRPush('a','B')

# Display the data of 0~(-1) from the left of array.
> redisLRange('a',0,-1)
[[1]]
[1] 2
[[2]]
[1] 1
[[3]]
[1] "A"
[[4]]
[1] "B"

# Pop a data from the right of data.
> redisRPop('a')
```

6.3.3.4 Operation of Class Set

```
> redisSAdd('A',runif(2))
> redisSAdd('A',55)

# Display the number of elements of object.
> redisSCard('A')
    [1] 2

# Display the set object elements in list.
> redisSMembers('A')
[[1]]
[1] 55
[[2]]
[1] 0.6494041 0.3181108
```

```
> redisSAdd('B',55)
> redisSAdd('B',rnorm(3))

# Display the number of elements of object.
> redisSCard('B')
[1] 2

# Display the set object elements in list.
> redisSMembers('B')
[[1]]
[1] 55
[[2]]
[1]  0.1074787 1.3111006 0.8223434

# Different set.
> redisSDiff(c('A','B'))
[[1]]
[1]  0.6494041 0.3181108

# Intersection.
> redisSInter(c('A','B'))
[[1]]
[1] 55

# Union.
> redisSUnion(c('A','B'))
[[1]]
[1] 55
[[2]]
[1]  0.1074787 1.3111006 0.8223434
[[3]]
[1]  0.6494041 0.3181108
```

6.3.3.5 Interaction between rredis and Redis-cli

Insert data from Redis client and read data using rredis.

```
# Open redis client.
~ redis-cli
redis 127.0.0.1:6379> set shell "Greetings, R client!"
OK

# Read data.
> redisGet('shell')
[1] "Greetings, R client!"
```

Use rredis to insert data and use Redis client to read data.

```
# Insert data.
> redisSet('R', 'Greetings, shell client!')
[1] "OK"

# Read data(gibberish).
redis 127.0.0.1:6379> get R
"X\\x00\x00\x00\x02\x00\x02\x0f\x00\x00\x02\x03\x00\x00\x00\x00\x10\
x00\x00\x00\x01\x00\x04\x00\\x00\x00\x00\x18Greetings, shell
client!"
```

Solve the problem of gibberish by transferring the data in array of Raw(charToRaw).

```
> redisSet('R', charToRaw('Greetings, shell client!'))
[1] TRUE

# Load the data normally.
redis 127.0.0.1:6379> get R
"Greetings, shell client!"
```

6.3.4 Test Case of rredis

As we've demonstrated the use of rredis, now let's run a test case. Read a data file. The file includes three items, user id, password and E-mail from left to right. Create a suitable data model in Redis for storage and query. First, define the data model.

KEY:
  – Users: user id
VALUE:
  – Id: user id
  – Pw: password
  – Email: E-mail address

Data file: data5.txt.

```
wolys # wolysopen111 # wolys@21cn.com
coralshanshan # 601601601 # zss1984@126.com
pengfeihuchao # woaidami # 294522652@qq.com
simulategirl # @#$9608125 # simulateboy@163.com
daisypp # 12345678 # zhoushigang_123@163.com
sirenxing424 # tfiloveyou # sirenxing424@126.com
raininglxy # 1901061139 # lixinyu23@qq.com
leochenlei # leichenlei # chenlei1201@gmail.com
z370433835 # lkp145566 # 370433835@qq.com
cxx0409 # 12345678 # cxx0409@126.com
xldq_1 # 06122211 # viv093@sina.com
```

Then, read the data file to memory. Create a Redis connection, insert the data to Redis in the form of loop, and output corresponding VALVE value using sers:wolys as KEY.

```
# Read data.
> data<-scan(file="data5.txt",what=character(),sep=" ")
> data<-data[which(data!='#')]

> data
[1] "wolys"            "wolysopen111"     "wolys@21cn.com"
[4] "coralshanshan"    "601601601"        "zss1984@126.com"
[7] "pengfeihuchao"    "woaidami"         "294522652@qq.com"
[10] "simulategirl"    "@#$9608125"       "simulateboy@163.
com"
[13] "daisypp"         "12345678"         "zhoushigang_123@163.
com"
[16] "sirenxing424"    "tfiloveyou"       "sirenxing424@126.
com"
[19] "raininglxy"      "1901061139"       "lixinyu23@qq.com"
[22] "leochenlei"      "leichenlei"       "chenlei1201@gmail.
com"
[25] "z370433835"      "lkp145566"        "370433835@qq.com"
[28] "cxx0409"         "12345678"         "cxx0409@126.com"
[31] "xldq_l"          "0612221l"         "viv093@sina.com"

# Connect redis.
> redisConnect(host="192.168.1.101",port=6379)
> redisFlushAll()
> redisKeys()

# Insert data in loop.
> id<-NULL
> for(i in 1:length(data)){
+   if(i %% 3 == 1) {
+     id<-data[i]
+     redisSAdd(paste("users:",id,sep=""),paste("id:",id,sep=""))
+   } else if(i %% 3 == 2) {
+     redisSAdd(paste("users:",id,sep=""),paste("pw:",data[i],sep=""))
+   } else {
+     redisSAdd(paste("users:",id,sep=""),paste("email:",data[i],
sep=""))
+   }
+}

# List all the KEY.
> redisKeys()
[1] "users:cxx0409"       "users:sirenxing424"  "users:simulategirl"
"users:xldq_l"
[5] "users:coralshanshan" "users:raininglxy"    "users:pengfeihuchao"
"users:leochenlei"
[9] "users:daisypp"       "users:wolys"         "users:z370433835"

# Query VALUE through KEY.
> redisSMembers("users:wolys")
```

```
[[1]]
[1] "pw:wolysopen111"

[[2]]
[1] "email:wolys@21cn.com"

[[3]]
[1] "id:wolys"

# Close redis connection.
> redisClose()
```

Thus we've finished the whole test case. Redis is a very efficient in-memory database. By combining R with Redis, we can construct a powerful real-time computing application.

6.4 Connecting R with NoSQL: Cassandra

Question

How do we use R to connect Cassandra?

Cassandra is an open source distributed database management system designed to handle large amounts of data across many commodity servers. Some techniques of Cassandra, including multiple datacenters, consistent hashing, and BlomFilter, provide new design ideas for the following NoSQL products. This section introduces RCassandra to connect R with Cassandra.

6.4.1 Environment Preparation of Cassandra

The name Cassandra comes from Greek mythology. It's the name of a tragic prophetess of Troy, so the logo of this project is a sparkling eye. The data of Cassandra will be written in multiple nodes to guarantee the reliability of data. In the compromise of consistency, availability, and partition tolerance (CAP), Cassandra is quite flexible. It allows users the options of demanding all the copies be consistent (high consistency), or one copy would be enough (high availability), or most copies would need to be consistent decided by vote (compromise) in reading. Thus Cassandra is suitable in all the nodes, in the environment of network failure or in the setting of a multidata center.

For environment preparation, I choose Linux Ubuntu here. You may choose other types of Linux according to your preference. The server environment of Cassandra is:

- Linux Ubuntu 12.04.2 LTS 64bit server
- Java JDK 1.6.0_45
- Cassandra 1.2.15

For information on the installation and configuration of Cassandra, please refer to Appendix E. Start the Cassandra server and use the command bin/cassandra.

```
# Enter the installation directory of Cassandra.
~cd /home/conan/tookit/cassandra1215

# Start Cassandra.
~ bin/cassandra

# View the system process.
~ ps -aux|grep cassandra

# View the starting log.
~ cat /var/log/cassandra/system.log |less
INFO [main] 2014-03-22 06:28:08,777 CassandraDaemon.java (line 119)
Logging initialized
INFO [main] 2014-03-22 06:28:08,795 CassandraDaemon.java (line 144)
JVM vendor/version: Java HotSpot(TM) 64-Bit Server VM/1.6.0_45
INFO [main] 2014-03-22 06:28:08,799 CassandraDaemon.java (line 182)
Heap size: 2051014656/2051014656
```

6.4.2 Function Library of RCassandra

RCassandra supports only 17 functions. Compared to rredis and rmongodb, the functions of RCassandra are relatively fewer. The 17 functions of RCassandra cannot even cover all the operations of Cassandra. In this case, some of the basic operations do not have the support of functions, so we need to handle those in the command line. I hope that RCassandra can be further developed to improve its function.

6.4.2.1 17 Functions

The following are the 17 functions and their comparison to the command of Cassandra.

```
RC.close                  RC.insert
RC.cluster.name           RC.login
RC.connect                RC.mget.range
RC.consistency            RC.mutate
RC.describe.keyspace      RC.read.table
RC.describe.keyspaces     RC.use
RC.get                    RC.version
RC.get.range              RC.write.table
RC.get.range.slices
```

6.4.2.2 Comparison of Basic Operations between Cassandra and RCassandra

1. Connect to cluster.

```
Cassandra:
connect 192.168.1.200/9160;

RCassandra:
conn<-RC.connect(host="192.168.1.200",port=9160)
```

2. View the name of the current cluster.

```
Cassandra:
show cluster name;

RCassandra:
RC.cluster.name(conn)
```

3. List all the keyspaces of the current cluster.

```
Cassandra:
show keyspaces;

RCassandra:
RC.describe.keyspaces(conn)
```

4. View the keyspace of DEMO.

```
Cassandra:
show schema DEMO;

RCassandra:
RC.describe.keyspace(conn,'DEMO')
```

5. Select the keyspace of DEMO.

```
Cassandra:
use DEMO;

RCassandra:
RC.use(conn,'DEMO')
```

6. Set the consistency level.

```
Cassandra:
consistencylevel as ONE;

RCassandra:
RC.consistency(conn,level="one")
```

7. Insert data.

```
Cassandra:
set Users[1][name] = scott;

RCassandra:
RC.insert(conn,'Users','1', 'name', 'scott')
```

8. Insert database.

```
Cassandra:
NA

RCassandra:
RC.write.table(conn, "Users", df)
```

9. Read all the data of the column family.

```
Cassandra:
list Users;

RCassandra:
RC.read.table(conn,"Users")
```

10. Read data.

```
Cassandra:
get Users[1]['name'];

RCassandra:
RC.get(conn,'Users','1', c('name'))
```

11. Quit the connection.

```
Cassandra:
exit; quit;

RCassandra:
RC.close(conn)
```

6.4.3 Basic Operations of RCassandra

We use RCassandra for all the basic function operations and take the iris data set in R as an example to introduce how to use RCassandra to operate the Cassandra database. Use the RCassandra client of R to access the Cassandra server remotely for the test.

Client environment:

■ Linux: Ubuntu 12.04.2 LTS 64bit server
■ R: 3.0.1, x86_64-pc-linux-gnu (64-bit)

First, start R, install RCassandra package and create a server connection.

```
# Start R.
~ R

# Install RCassandra.
> install.packages('RCassandra')

# Load RCassandra class library.
> library(RCassandra)
# Create server connection.
> conn<-RC.connect(host="192.168.1.200")

# The name of current cluster(the name of cluster of 2 nodes).
> RC.cluster.name(conn)
[1] "case1"

# Version of current protocol.
> RC.version(conn)
[1] "19.36.0"
```

Keyspaces operation:

```
# List all the configuration information of keyspaces.
> RC.describe.keyspaces(conn)

# List all the configuration information of keyspaces named DEMO.
> RC.describe.keyspace(conn, "DEMO")
```

Then there is the operation of data. We should note that we cannot create a column family in RCassandra, so we need to create a column family through Cassandra commands.

```
# Insert iris data.
> head(iris)
  Sepal.Length Sepal.Width Petal.Length Petal.Width Species
1          5.1         3.5          1.4         0.2  setosa
2          4.9         3.0          1.4         0.2  setosa
3          4.7         3.2          1.3         0.2  setosa
4          4.6         3.1          1.5         0.2  setosa
5          5.0         3.6          1.4         0.2  setosa
6          5.4         3.9          1.7         0.4  setosa

# iris is a data.frame.
> RC.write.table(conn, "iris", iris)
attr(,"class")
[1] "CassandraConnection"

# View the value of the first row, Sepal.Length column.
> RC.get(conn, "iris", "1", c("Sepal.Length", "Species"))
          key  value              ts
1 Sepal.Length   5.1 1.372881e+15
2 Species setosa     1.372881e+15
# Note: ts is timestamp.

# View the first row.
> RC.get.range(conn, "iris", "1")
          key  value              ts
1 Petal.Length    1.4 1.372881e+15
2 Petal.Width     0.2 1.372881e+15
3 Sepal.Length    5.1 1.372881e+15
4 Sepal.Width     3.5 1.372881e+15
5 Species setosa     1.372881e+15

# View the data set.
> r <- RC.get.range.slices(conn, "iris")
class(r)
[1] "list"

> r[[1]]
          key  value              ts
1 Petal.Length    1.7 1.372881e+15
2 Petal.Width     0.4 1.372881e+15
3 Sepal.Length    5.4 1.372881e+15
4 Sepal.Width     3.9 1.372881e+15
5 Species setosa     1.372881e+15

> rk <- RC.get.range.slices(conn, "iris", limit=0)
> y <- RC.read.table(conn, "iris")
> y <- y[order(as.integer(row.names(y))),]
```

```
> head(y)
  Petal.Length Petal.Width Sepal.Length Sepal.Width Species
1          1.4         0.2          5.1         3.5  setosa
2          1.4         0.2          4.9         3.0  setosa
3          1.3         0.2          4.7         3.2  setosa
4          1.5         0.2          4.6         3.1  setosa
5          1.4         0.2          5.0         3.6  setosa
```

Some commonly used operations that are not supported:

■ Create keyspaces, delete keyspaces.
■ Create column family, delete column family.
■ Delete a certain row.
■ Delete the data of certain columns in a certain row.

6.4.4 A Case Using RCassandra

Here is a case of business demand to help us deepen our understanding of RCassandra. This is a very simple business scene. The demand is as follows: (1) Create a Users column family, including two rows: name, password. (2) Add a new column age when data already exist. First let's create a Users column family in the command line.

```
[default@DEMO] create column family Users
...      with key_validation_class = 'UTF8Type'
...      and comparator = 'UTF8Type'
...      and default_validation_class = 'UTF8Type';
89a2fb75-f7d0-399e-b017-30a974b19f4a
```

Insert data in RCassandra, including two columns: name and password.

```
> df<-data.frame(name=c('a1','a2'),password=c('a1','a2')) >
print(df)
  name password
1   a1       a1
2   a2       a2

# Insert data.
> RC.write.table(conn, "Users", df)
attr(,"class")
[1] "CassandraConnection"

# View the data.
> RC.read.table(conn,"Users")
    name password
2     a2       a2
1     a1       a1
```

Insert a new line KEY = 1234, and add column age.

```
> RC.insert(conn,'Users','1234', 'name', 'scott')
> RC.insert(conn,'Users','1234', 'password', 'tiger')
> RC.insert(conn,'Users','1234', 'age', '20')

# View the data.
> RC.read.table(conn,"Users")
     age  name password
1234  20 scott    tiger
2     NA    a2       a2
1     NA    a1       a1
```

Modify the row of KEY = 1: name = a11, age = 12.

```
> RC.insert(conn,'Users','1', 'name', 'a11')
> RC.insert(conn,'Users','1', 'age', '12')

# View the data.
> RC.read.table(conn,"Users")
     age  name password
1234  20 scott    tiger
2     NA    a2       a2
1     12   a11       a1
```

6.4.5 Decline of Cassandra

As Hadoop rises and HBase, a family product of Hadoop, is becoming widely used, more and more applications constructed on Cassandra are now transferred to HBase. We may find some of the reasons for the decline of Cassandra on a technical level.

1. Poor performance in reading
   The decentralized design determines that Cassandra runs calculations through anti-entropy. This will cost performance greatly, and even severely affect the running of the server.
2. Slow data synchronization (final consistency delay may be very large)
   Decentralized design needs all the nodes to pass information and send a notice informing of the status. If there are multiple replica sets and circumstances of nodes going down, the delay of data consistency would be very large and the efficiency would be very low.
3. It is very inflexible to replace query with insert and update. All the queries demand to be defined in advance.
   In contrast with the fact that many databases choose to optimize for reading, Cassandra pays great attention in optimizing writing. So it's very suitable for streaming data storage, especially for those whose writing loads are higher than their reading load. Compared with HBase, Cassandra is higher in random access performance, but not as good in interval scanning. So Cassandra can serve as a real-time query cache of HBase, where HBase runs a big data process in batch and Cassandra provides an interface for random access.

4. Cassandra doesn't support Hadoop and can't implement MapReduce.

Today, Hadoop has become the synonym for big data. As a frame of big data, Cassandra will be replaced one day if it continues its condition of not supporting Hadoop and MapReduce unless it positions itself in a different area or provides a solution for big data. DataStax is now reconstructing the file system of HDFS using Cassandra. We don't know whether they can succeed, but let's expect the further development of Cassandra!

6.5 Connecting R with NoSQL: Hive

Question

How do we connect Hive using R?

Hive is a program interface of Hadoop. It adopts SQL-like syntax, which makes it a relatively easy task for a data analyst to master it quickly. Hive has made the world of Java simpler and lighter and helps Hadoop to be accepted by nonprogrammers. Starting with Hive, an analyst can master big data too. This section introduces how to use RHive to help R connect Hive.

6.5.1 Environment Preparation of Hive

Hive is a data warehouse infrastructure based on Hadoop, which provides a series of tools for the extract, transform, and load of data (ETL). ETL is a big data mechanism for storing, querying, and analyzing data in Hadoop. Hive has defined a simple SQL-like query language, called HQL. It allows users familiar SQL to query data. Hive doesn't have a specific format of data, so it can be worked well on Thrift. Hive controls separator and allows users to assign data format.

The difference between Hive and relational databases are as follows:

- Different data storage: Hive is based on HDFS of Hadoop, while relational databases are based on a local file system.
- Different computing model: Hive is based on the MapReduce of Hadoop, while relational databases are based on a memory computing model of indexes.
- Different application scene: The OLAP data warehouse system provides a big data query for Hive, with a poor real-time performance, while the OLTP transactional system serves relational databases in real-time query businesses.

■ Different expansibility: It's very easy to increase distributional storage capacity and computing power in Hive as it is based on Hadoop; and it's rather difficult to make horizontal scaling in relational databases. We need to increase serial performance consistently.

Hive is a data warehouse product based on Hadoop, so we need to have a Hadoop environment. For the installation and configuration of Hadoop, please refer to Appendix G. For environment preparation, I choose Linux Ubuntu here. You may choose other types of Linux according to your preference.

■ Linux Ubuntu 12.04.2 LTS 64bit server
■ Java JDK 1.6.0_45
■ Hadoop 1.1.2
■ hive 0.9.0
■ IP: 192.168.1.210

```
# Start hiveserver serve in the background through nohup.
~ nohup hive --service hiveserver  &
Starting Hive Thrift Server
```

After Hive is installed, start the serve of hiveserver.
Simple operation of Hive Shell:

```
# Open hive shell.
~ hive shell
Logging initialized using configuration in file:/home/conan/hadoop/
hive-0.9.0/conf/hive-log4j.properties
Hive history file=/tmp/conan/hive_job_log_
conan_201306261459_153868095.txt

# View the tables of hive.
hive> show tables;
hive_algo_t_account
o_account
r_t_account
Time taken: 2.12 seconds

# View the data of table o_account.
hive> select * from o_account;
1       abc@163.com     2013-04-22 12:21:39
2       dedac@163.com   2013-04-22 12:21:39
3       qq8fed@163.com  2013-04-22 12:21:39
4       qw1@163.com     2013-04-22 12:21:39
5       af3d@163.com    2013-04-22 12:21:39
6       ab34@163.com    2013-04-22 12:21:39
```

```
7           q8d1@gmail.com   2013-04-23 09:21:24
8           conan@gmail.com  2013-04-23 09:21:24
9           adeg@sohu.com    2013-04-23 09:21:24
10          ade121@sohu.com  2013-04-23 09:21:24
11          addde@sohu.com   2013-04-23 09:21:24
Time taken: 0.469 seconds
```

6.5.2 Installation of RHive

Then, let's install RHive in a computer with a Hive environment. The client environment is as follows.

- Linux: Ubuntu 12.04.2 LTS 64bit server
- R: 3.0.1, x86_64-pc-linux-gnu (64-bit)

Note: RHive only supports Linux.
First, install the dependent Library of R, rJava.

```
# Recommend that root privilege be used to install rJava.
~ sudo R CMD javareconf

# Start R.
~ sudo R
> install.packages("rJava")

# Install RHive.
> install.packages("RHive")

# Load RHive.
> library(RHive)
Loading required package: rJava
Loading required package: Rserve
This is RHive 0.0-7. For overview type '?RHive'.
HIVE_HOME=/home/conan/hadoop/hive-0.9.0
call rhive.init() because HIVE_HOME is set.
```

When loading RHive, the environment variables of local Hive will be loaded in the environment of R automatically.

6.5.3 Function Library of RHive

RHive provides 52 functions, corresponding to different operations of Hive. We will not list the entire RHive function library but only select some commonly used functions to compare with the operation of Hive. If you are interested in other functions, you may check the official documents of RHive. Here is the comparison of basic operations between RHive and Hive.

1. Connect to Hive.

```
Hive:
hive shell

RHive:
rhive.connect("192.168.1.210")
```

2. List all the tables of Hive.

```
Hive:
show tables;

RHive:
rhive.list.tables()
```

View the table structure.

```
Hive:
desc o_account;

RHive:
rhive.desc.table('o_account')
rhive.desc.table('o_account',TRUE)
```

Run the HQL query.

```
Hive:
select * from o_account;

RHive:
rhive.query('select * from o_account')
```

View the HDFS directory.

```
Hive:
dfs -ls /;

RHive:
rhive.hdfs.ls()
```

View the file contents of HDFS.

```
Hive:
dfs -cat /user/hive/warehouse/o_account/part-m-00000;

RHive:
rhive.hdfs.cat('/user/hive/warehouse/o_account/part-m-00000')
```

Disconnect.

```
Hive:
quit;

RHive:
rhive.close()
```

6.5.4 Basic Operations of RHive

```
# Initialization.
> rhive.init()

# Connect to hive.
> rhive.connect("192.168.1.210")

# View all the tables.
> rhive.list.tables()
             tab_name
1 hive_algo_t_account
2           o_account
3         r_t_account

# View table structure.
> rhive.desc.table('o_account');
    col_name data_type comment
1         id       int
2      email    string
3 create_date    string

# Run HQL query.
> rhive.query("select * from o_account");
   id          email         create_date
1   1     abc@163.com 2013-04-22 12:21:39
2   2   dedac@163.com 2013-04-22 12:21:39
3   3 qq8fed@163.com 2013-04-22 12:21:39
4   4     qw1@163.com 2013-04-22 12:21:39
```

```
5   5     af3d@163.com 2013-04-22 12:21:39
6   6     ab34@163.com 2013-04-22 12:21:39
7   7  q8d1@gmail.com 2013-04-23 09:21:24
8   8 conan@gmail.com 2013-04-23 09:21:24
9   9    adeg@sohu.com 2013-04-23 09:21:24
10 10 ade121@sohu.com 2013-04-23 09:21:24
11 11  addde@sohu.com 2013-04-23 09:21:24

# Close connection.
> rhive.close()
[1] TRUE

# Create temporary tables.
> rhive.block.sample('o_account', subset="id<5")
[1] "rhive_sblk_1372238856"

> rhive.query("select * from rhive_sblk_1372238856");
  id         email          create_date
1  1      abc@163.com 2013-04-22 12:21:39
2  2    dedac@163.com 2013-04-22 12:21:39
3  3  qq8fed@163.com 2013-04-22 12:21:39
4  4      qw1@163.com 2013-04-22 12:21:39

# View hdfs files.
> rhive.hdfs.ls('/user/hive/warehouse/rhive_sblk_1372238856/')
  permission owner group length modify-time
1  rw-r--r-- conan supergroup    141 2013-06-26 17:28
                                                 file
1 /user/hive/warehouse/rhive_sblk_1372238856/000000_0

rhive.hdfs.cat('/user/hive/warehouse/
rhive_sblk_1372238856/000000_0')
1abc@163.com2013-04-22 12:21:39
2dedac@163.com2013-04-22 12:21:39
3qq8fed@163.com2013-04-22 12:21:39
4qw1@163.com2013-04-22 12:21:39

# Separate field data according to range.
> rhive.basic.cut('o_account','id',breaks='0:100:3')
[1] "rhive_result_20130626173626"
attr(,"result:size")
[1] 443

> rhive.query("select * from rhive_result_20130626173626");
          email          create_date     id
1      abc@163.com 2013-04-22 12:21:39 (0,3]
2    dedac@163.com 2013-04-22 12:21:39 (0,3]
3  qq8fed@163.com 2013-04-22 12:21:39 (0,3]
4      qw1@163.com 2013-04-22 12:21:39 (3,6]
5     af3d@163.com 2013-04-22 12:21:39 (3,6]
6     ab34@163.com 2013-04-22 12:21:39 (3,6]
7  q8d1@gmail.com 2013-04-23 09:21:24 (6,9]
```

```
 8   conan@gmail.com 2013-04-23 09:21:24  (6,9]
 9     adeg@sohu.com 2013-04-23 09:21:24  (6,9]
10  ade121@sohu.com 2013-04-23 09:21:24  (9,12]
11   addde@sohu.com 2013-04-23 09:21:24  (9,12]
```

Operate HDFS through Hive.

```
# View the file directory of hdfs.
> rhive.hdfs.ls()
  permission owner       group length       modify-time   file
1  rwxr-xr-x conan supergroup        0 2013-04-24 01:52 /hbase
2  rwxr-xr-x conan supergroup        0 2013-06-23 10:59 /home
3  rwxr-xr-x conan supergroup        0 2013-06-26 11:18 /rhive
4  rwxr-xr-x conan supergroup        0 2013-06-23 13:27 /tmp
5  rwxr-xr-x conan supergroup        0 2013-04-24 19:28 /user

# View the file contents of hdfs.
> rhive.hdfs.cat('/user/hive/warehouse/o_account/part-m-00000')
1abc@163.com2013-04-22 12:21:39
2dedac@163.com2013-04-22 12:21:39
3qq8fed@163.com2013-04-22 12:21:39
```

We've now connected the data channel between R and Hive using RHive and started to achieve big data operation based on the Hive system using R. In the next section, I introduce a RHive case of big data within a financial area.

6.6 Extract Reverse Repurchase Information from Historical Data Using RHive

Question

How do we perform financial data analysis using RHive?

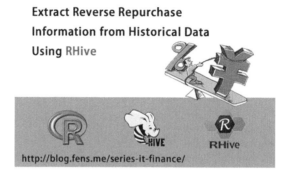

I've been in the area of finance for a relatively short time, and reverse repurchase is the first product that I've operated. Reverse repurchase is possibly a new word for most of us, or even for the experienced stock market participant. China's banking system experienced a severe lack of liquidity in July 2013. The overnight interbanking borrowing rate reached 30% at that time. When the banking system lacks liquidity and short-term interest rates rise much higher, other financial institutions sell their stocks and bonds to lend money to the banks.

6.6.1 Introduction of Reverse Repurchase

To put it simply, pledge-style bond repurchase is a short-term capital lending transaction in which both sides use bonds as pledges. The bond holder (repurchase party) pledges his bonds to get the right to use capital and pays a certain amount of interest in an appointed time to repurchase the bonds. And the capital holder (reverse repurchase party) is the transaction counterparty. In real practice, the bond is pledged to a third party institution, which makes the transaction safer and more convenient.

Bonds that can be repurchased include national bonds, most of the corporate bonds, and the bond part of the separate bonds. The stock exchange will publish the discount rate of the bonds that can be repurchased regularly. The discount rate refers to the ratio of the money that bond holders can get to the face value of bonds.

6.6.2 Storage Structure of Historical Data

System environment used in this section:

- Linux: Ubuntu 12.04.2 LTS 64bit
- R: 3.0.1 x86_64-pc-linux-gnu
- Domain name of the internal network: c1.wtmart.com

Table structure in Hive:

```
> rhive.desc.table('t_reverse_repurchase')
     col_name data_type
1   tradedate    string
2   tradetime    string
3  securityid    string
4      bidpx1    double
5     bidsize1    double
6     offerpx1    double
7   offersize1    double
```

6.6.3 Extract Data Using RHive

Log in the c1.wtmart.com server and open R client.

```
# Start R.
~ R

# Load RHive.
> library(RHive)

# Initialization.
> rhive.init()
```

```
# Connect to Hive cluster.
> rhive.connect("c1.wtmart.com")
SLF4J: Class path contains multiple SLF4J bindings.
SLF4J: Found binding in [jar:file:/home/cos/toolkit/hive-0.9.0/lib/
slf4j-log4j12-1.6.1.jar!/org/slf4j/impl/StaticLoggerBinder.class]
SLF4J: Found binding in [jar:file:/home/cos/toolkit/hadoop-1.0.3/
lib/slf4j-log4j12-1.4.3.jar!/org/slf4j/impl/StaticLoggerBinder.
class]
SLF4J: See http://www.slf4j.org/codes.html#multiple_bindings for an
explanation.

# View the current table.
> rhive.list.tables()
              tab_name
1            t_hft_day    // Historical data table.
2            t_hft_tmp    // Temporary table.
4 t_reverse_repurchase    // Reverse repurchase table.
```

View the historical data fragmentation of all stocks: the test data are from June 27, 2013 to July 26, 2014.

```
# View the fragmentation of table t_hft_day.
> rhive.query("SHOW PARTITIONS t_hft_day");
             partition
1   tradedate=20130627
2   tradedate=20130628
3   tradedate=20130701
4   tradedate=20130702
5   tradedate=20130703
6   tradedate=20130704
7   tradedate=20130705
8   tradedate=20130708
9   tradedate=20130709
10  tradedate=20130710
11  tradedate=20130712
12  tradedate=20130715
13  tradedate=20130716
14  tradedate=20130719
15  tradedate=20130722
16  tradedate=20130723
17  tradedate=20130724
18  tradedate=20130725
19  tradedate=20130726
```

Extract the data of "One Day Reverse Repurchase of Shanghai Stock Exchange" (204001) and "One Day Reverse Repurchase of Shenzhen Stock Exchange" (131810).

```
# Delete reverse repurchase table.
> rhive.drop.table("t_reverse_repurchase")

# Create reverse repurchase table and insert data.
> rhive.query("CREATE TABLE t_reverse_repurchase AS SELECT tradedate
,tradetime,securityid,bidpx1,bidsize1,offerpx1,offersize1 FROM t_
hft_day where tradedate>=20130722 and securityid in
(131810,204001)");

# View the data result set.
> rhive.query("SELECT securityid,count(1) FROM t_reverse_repurchase
group by securityid");
  securityid  X_c1
1     131810 17061
2     204001 12441
```

Load it to the memory of R.

```
# Query data from Hive, and assign the value to bidpx1 in R.
> bidpx1<-rhive.query("SELECT securityid,concat(tradedate,tradetime)
as tradetime,bidpx1 FROM t_reverse_repurchase"); #查看记录条数 >
nrow(bidpx1)
[1] 29502

# View the data.
> head(bidpx1)
  securityid      tradetime bidpx1
1     131810 20130724145004  2.620
2     131810 20130724145101  2.860
3     131810 20130724145128  2.850
4     131810 20130724145143  2.603
5     131810 20130724144831  2.890
6     131810 20130724145222  2.600
```

Use ggplot2 draw the trend of these two products in one week, as in Figure 6.1.

```
# Load ggplot2.
> library(ggplot2)
> g<-ggplot(data=bidpx1, aes(x=as.POSIXct(tradetime,format="%Y%m%d%H
%M%S"), y=bidpx1))
> g<-g+geom_line(aes(group=securityid,colour=securityid))
> g<-g+xlab('tradetime')+ylab('bidpx1')

# Output the image to file.
> ggsave(g,file="01.png",width=12,height=8)
```

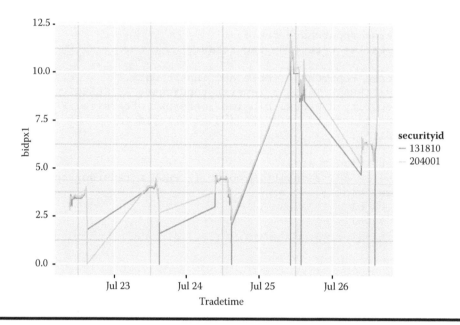

Figure 6.1   Trend in one week.

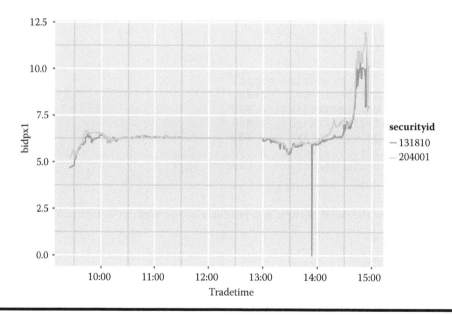

Figure 6.2   Trend in one day.

Then the trend in one day, July 26, 2013, is as in Figure 6.2.

```
> bidpx1<-rhive.query("SELECT securityid,concat(tradedate,tradetime)
as tradetime,bidpx1 FROM t_reverse_repurchase WHERE
tradedate=20130726");
> g<-ggplot(data=bidpx1, aes(x=as.POSIXct(tradetime,format="%Y%m%d%H
%M%S"), y=bidpx1))
> g<-g+geom_line(aes(group=securityid,colour=securityid))
> g<-g+xlab('tradetime')+ylab('bidpx1')
> ggsave(g,file="02.png",width=12,height=8)
```

6.6.4 Strategy Model and Its Implementation

Print out the two images of the two curves. We can see that 131810 has always followed the trend 204001 and was lower than 204001 most of the time. Here is a simple strategy analysis to decide when to sell 131810 according to the change of 204001.

1. Standardize 131810 and 204001 to each minute.
2. Take it as selling signal when 131810 and 204001 cross.
3. When the offer1 price of the latter cross of 131810 and 204001 is 10% higher than the offer1 price of the former cross, set the latter cross as a local optimal sell signal.

Extract the data of 131810 and 204001, and store them to table t_reverse_repurchase.

```
# Sign in R.
> library(RHive)
> rhive.init()
> rhive.connect("c1.wtmart.com")

# Extract data of 131810 and 204001.
> rhive.drop.table("t_reverse_repurchase")
> rhive.query("CREATE TABLE t_reverse_repurchase AS SELECT tradedate
,tradetime,securityid,bidpx1,bidsize1,offerpx1,offersize1 FROM t_
hft_day where securityid in (131810,204001)");

# View the data set.
> rhive.query("select count(1),tradedate from t_reverse_repurchase
group by tradedate")
    X_c0 tradedate
1   4840  20130627
2   4792  20130628
```

```
 3  4677   20130701
 4  3124   20130702
 5  2328   20130703
 6  3787   20130704
 7  4294   20130705
 8  4977   20130708
 9  4568   20130709
10  6619   20130710
11  5633   20130712
12  6159   20130715
13  5918   20130716
14  6200   20130719
15  6074   20130722
16  5991   20130723
17  5899   20130724
18  5346   20130725
19  6192   20130726
```

Acquire the data of one day and make ETL. First load the package:

```
> library(ggplot2)
> library(scales)
> library(plyr)
```

Load the data of one week to memory:

```
> bidpx1<-rhive.query(paste("SELECT
securityid,tradedate,tradetime,bidpx1 FROM t_reverse_repurchase
WHERE tradedate>=20130722"));
```

Load the data of one day and make ETL:

```
> oneDay<-function(date){
+   d1<-bidpx1[which(bidpx1$tradedate==date),]
+   d1$tradetime2<-round(as.numeric(as.character(d1$tradetime))/100)*100
+   d1$tradetime2[which(d1$tradetime2<100000)] <- paste(0,d1$tradetime2
[which(d1$tradetime2<100000)],sep="")
+   d1$tradetime2[which(d1$tradetime2=='1e+05')]='100000'
+   d1$tradetime2[which(d1$tradetime2=='096000')]='100000'
+   d1$tradetime2[which(d1$tradetime2=='106000')]='110000'
+   d1$tradetime2[which(d1$tradetime2=='126000')]='130000'
+   d1$tradetime2[which(d1$tradetime2=='136000')]='140000'
+   d1$tradetime2[which(d1$tradetime2=='146000')]='150000'
+   d1
+ }
```

Standardize the data through averages:

```
> meanScale<-function(d1){
+   ddply(d1, .(securityid,tradetime2), summarize, bidpx1=mean(bidpx1))
+ }

> findPoint<-function(a1,a2){
+   bigger_point<-function(a1,a2){
+     idx<-c()
+     for(i in intersect(a1$tradetime2,a2$tradetime2)){
+       i1<-which(a1$tradetime2==i)
+       i2<-which(a2$tradetime2==i)
+       if(a1$bidpx1[i1]-a2$bidpx1[i2]>=-0.02){
+         idx<-c(idx,i1)
+       }
+     }
+     idx
+   }
+   remove_continuous_point<-function(idx){
+     idx[-which(idx-c(NA,rev(rev(idx)[-1]))==1)]
+   }
+   idx<-bigger_point(a1,a2)
+   remove_continuous_point(idx)
+ }


> bigger_point<-function(a1,a2){
+   idx<-c()
+   for(i in intersect(a1$tradetime2,a2$tradetime2)){
+     i1<-which(a1$tradetime2==i)
+     i2<-which(a2$tradetime2==i)
+     if(a1$bidpx1[i1]-a2$bidpx1[i2]>=-0.02){
+       idx<-c(idx,i1)
+     }
+   }
+   idx
+ }

> remove_continuous_point<-function(idx){
+   idx[-which(idx-c(NA,rev(rev(idx)[-1]))==1)]
+ }

+   idx<-bigger_point(a1,a2)
+   remove_continuous_point(idx)
+ }

> findOptimize<-function(d3){
+
    idx2<-which((d3$bidpx1-c(NA,rev(rev(d3$bidpx1)[-1])))/d3$bidpx1>0.1)
+   if(length(idx2)<1)
+     print("No Optimize point")
+   d3[idx2,]
+ }
```

```
> draw<-function(d2,d3,d4,date,png=FALSE){
+  g<-ggplot(data=d2, aes(x=strptime(paste(date,tradetime2,sep=""),
format="%Y%m%d%H%M%S"), y=bidpx1))
+  g<-g+geom_line(aes(group=securityid,colour=securityid))
+  g<-g+geom_point(data=d3,aes(size=1.5,colour=securityid))
+if(nrow(d4)>0){
+     g<-g+geom_text(data=d4,aes(label= format(d4$bidpx1,digits=4)),
colour="blue",hjust=0, vjust=0)
+  }
+  g<-g+xlab('tradetime')+ylab('bidpx1')
+  if(png){
+     ggsave(g,file=paste(date,".png",sep=""),width=12,height=8)
+  }else{
+     g
+  }
+ }

> date<-20130722
> d1<-oneDay(date)
> d2<-meanScale(d1)
> a1<-d2[which(d2$securityid==131810),]
> a2<-d2[which(d2$securityid==204001),]
> d3<-d2[findPoint(a1,a2),]
> d4<-findOptimize(d3)
> draw(d2,d3,d4,as.character(date),TRUE)
```

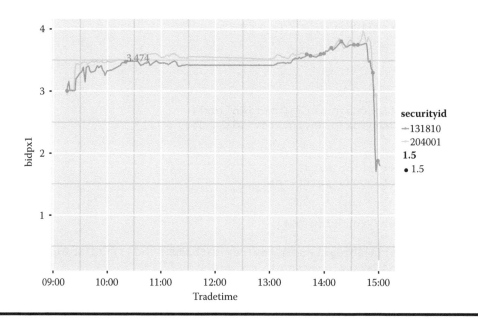

Figure 6.3   Strategy visualization on July 22, 2013.

Generate five images corresponding to five days of a week (Figures 6.3 to 6.7). The back-test of the data shows that the signals given by the programmatic strategy are all good selling points. Their performance is better than my manual operation.

Through the comparison of the data in a week, we find that this simple strategy can bring us some benefits. It's better than the man-made decision

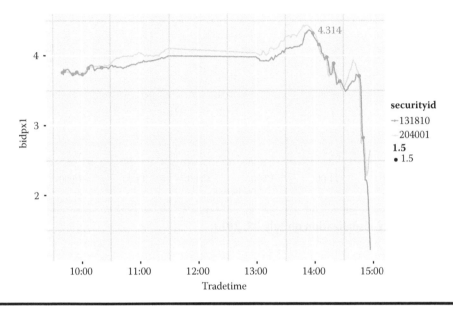

Figure 6.4   Strategy visualization on July 23, 2013.

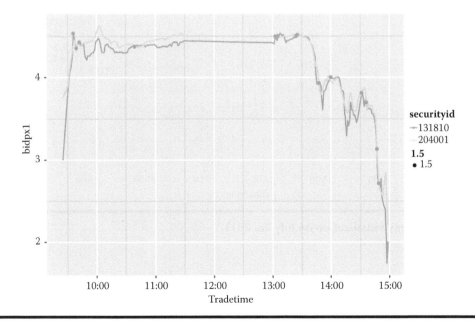

Figure 6.5   Strategy visualization on July 24, 2013.

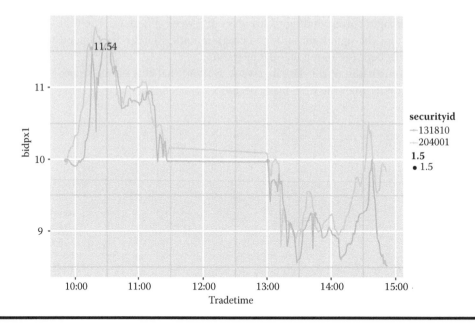

Figure 6.6   Strategy visualization on July 25, 2013.

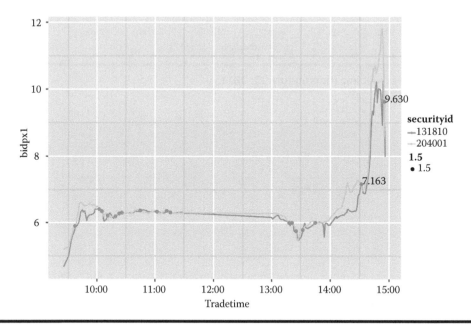

Figure 6.7   Strategy visualization on July 26, 2013.

Chapter 7

RHadoop

This chapter mainly introduces how to use R to access a Hadoop cluster through a RHadoop tool, help readers to manage HDFS using R, develop MapReduce programs, and access HBase access. This chapter uses R to implement MapReduce program cases based on a collaborative filtering algorithm. It's more concise than Java.

7.1 R Has Injected Statistical Elements into Hadoop

Question
Why should we combine R with Hadoop?

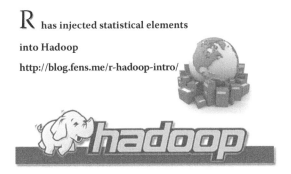

R and Hadoop belong to two different disciplines. They have different user groups, they are based on two different knowledge systems, and they do different things. But data, as their intersection, have made the combination with R and Hadoop an interdisciplinary choice, a tool to mine the value of data.

7.1.1 Introduction to Hadoop

For people in the IT world, Hadoop is a rather familiar technique. Hadoop is a distributed systematic basic construction, managed by Apache Fund. Users can develop distributional programs even if they know little about underlying details of distributed construction, and make full use of the high-speed calculation and storage of clusters. Hadoop has implemented a distributed file system (Hadoop Distributed File System, HDFS). HDFS is high in fault tolerance and designed to be deployed in low-cost hardware. In addition, it provides high throughput to access the data of applications, which makes it suitable for the statistical analysis of data warehouses.

Hadoop has many family members, including Hive, HBase, Zookeeper, Avro, Pig, Ambari, Sqoop, Mahout and Chukwa, and so forth. I'll provide a brief introduction to these.

- Hive is a data warehouse tool based on Hadoop. It can map the structured data file to a database table and quickly achieve simple MapReduce statistics through SQL-like statements. It is very suitable for statistical analysis for data warehouses because it doesn't need to develop special MapReduce applications.
- Pig is big data analysis tool based on Hadoop. It provides SQL-like language, Pig Latin. Pig Latin will transform the SQL-like data analysis request to a series of optimized MapReduce calculations.
- HBase is a highly reliable, high-performance, column-oriented, and scalable distributed storage system. We can construct massive structured storage clusters on a low-cost PC Server using HBase.
- Sqoop is a tool to transfer data between Hadoop and relational database. We can transfer the data of relational database(MySQL, Oracle, Postgres, etc.) to Hadoop and vice versa.
- Zookeeper is a distributed open source coordination service designed for distributed applications. It's mainly used for solving the data management problems occurring in distributed applications, simplifying the difficulty of coordination and management of distributed applications, and increasing the performance of a distributed service.
- Mahout is a distributed framework based on the machine learning and data mining of Hadoop. It uses MapReduce to achieve some of its data mining algorithm and solve the problem of parallel mining.
- Avro is a data serialization system designed for data-intensive applications with massive data exchange. Avro is a new data serialization format and transfer facility, and it will replace the original IPC mechanism of Hadoop.
- Ambari, based on the Web, supports the supply, management, and monitoring of Hadoop clusters.
- Chukwa is an open source data collection system used for monitoring large distributed systems. It can collect all kinds of data, transform them into a file suited to be processed by Hadoop, and store them in HDFS for Hadoop to perform MapReduce operations.

Hadoop started to develop independently with MapReduce and HDFS in 2006. The Hadoop family has now been incubating several top Apache projects. Especially in recent years, the development of Hadoop has grown much faster. It integrates many new techniques with it, such as YARN, Hcatalog, Oozie, and Cassandra, which makes it very hard to follow. For more introductory information, please refer to the series of articles in the author's blog about the Hadoop family.

7.1.2 Why Should We Combine R with Hadoop?

R and Hadoop are all-stars playing important roles in their own fields. When speaking of combining R with Hadoop, many software developers will look at it from the perspective of computing. Here are two questions.

- Question 1: Why should we combine R with Hadoop as the Hadoop family is already so powerful?
- Question 2: Mahout can also perform data mining and machine learning. So what's the difference between R and Mahout?

Here is my answer to these two questions.

7.1.2.1 Why Should We Combine R with Hadoop When the Hadoop Family Is Already So Powerful?

1. The power of the Hadoop family lies in its processing of big data. Hadoop makes it possible to perform the calculation of data at the GB, TB, or even PB level.
2. The power of R lies in its performance in statistical analysis. Before Hadoop was developed, the processing of big data included sampling, hypothesis testing, and regression analysis was almost exclusively performed by statisticians and R.
3. From the aforementioned two points, we may find that the focus of Hadoop is total data analysis, while the focus of R is sample data analysis. So combining the two techniques is very complementary!

Now, let's simulate a scene: analyze the access log of a news website of 1PB and predict the change in flow in the future. The specific process can be divided into four steps.

- Use R to perform data analysis, a construct regression model on the business object, and define indicators.
- Use Hadoop to extract total indicator data from the massive log data.
- Use R to verify and optimize the indicator data.
- Use the distributed algorithm of Hadoop to rewrite the R model and deploy it online.

In this scene, R and Hadoop both play important roles. If we adopt the thinking pattern of programmers and use only Hadoop in the whole process, there will be no data modeling and process of proving, and the forecasting results will also be problematic. If we adopt the thinking pattern of statisticians and use only R, the forecasting result of sampling will also be biased. So combining R with Hadoop is an inevitable orientation of the industry. The intersection of the industry with academia has provided infinite space for the imagination of researchers in interdisciplinary studies.

7.1.2.2 Mahout Can Also Perform Data Mining and Machine Learning. So What's the Difference between R and Mahout?

1. Mahout is an algorithm framework of data mining and machine learning based on Hadoop. The focus of Mahout is to solve the computing problems of big data.
2. An algorithm supported by Mahout includes collaborative filtering, recommended algorithm, clustering algorithm, sorting algorithm, linear discriminant analysis (LDA), naïve

Bayes, random forest, and so forth. Most of the algorithms in Mahout are based on distance. After matrix decomposition, they make full use of the parallel computing framework of MapReduce and accomplish their computing tasks efficiently.

3. Many of the data mining algorithms of Mahout cannot be paralleled by MapReduce. Besides, all the current models of Mahout are general computing models. If we apply these models in projects directly, the computing result will be only a little better than the random result. The secondary development of Mahout requires profound technical basis of Java and Hadoop, or even basic knowledge of linear algebra, probability statistics, and introduction to algorithms. So it's not easy to master Mahout.

4. R provides most of the algorithms supported by Mahout. It also provides many algorithms that are not supported by Mahout. In addition, the growth of algorithms in R is faster than in Mahout, and it's simple in development and flexible in parameter configuration. R is also very fast in small data cluster computing.

Though Mahout can also perform well in data mining and machine learning, its strong field doesn't overlap with R. Only by combining the advantages of all sides and choosing the right technique in suitable fields can we truly guarantee the quality of programs we make.

7.1.3 How to Combine Hadoop with R

From the preceding introduction we find that R and Hadoop are complementary. But in the application of enterprises, R and Hadoop are still used individually. The demand of the market will naturally guide some of the experts to fill in the blank. So how do we combine Hadoop with R? Here I give five ways of combining.

7.1.3.1 RHadoop

RHadoop is a product that combines Hadoop and R. It was developed by RevolutionAnalytics, and its code is published in Github. RHadoop includes three R packages (rmr, rhdfs, and rHBase), each corresponding to the three parts of Hadoop system construction: MapReduce, HDFS, and HBase. This chapter gives a detailed introduction to the installation and use of RHadoop.

7.1.3.2 RHive

RHive, developed by the Korean company NexR, is a tool package to access Hive directly through R. Sections 6.5 and 6.6 have provided information on the installation and use of RHive.

7.1.3.3 Rewrite Mahout

Using R to rewrite Mahout is also a way of combining Hadoop with R. I've made some attempts on this, and they will be introduced in the next book of this series, *R for Geeks: Advanced Development.*

7.1.3.4 Use Hadoop to Call R

All of the preceding paragraphs discuss how to use R to call Hadoop. Of course we can reverse the operation: connect Java with R and use Hadoop to call R.

I've written three examples of using Java to call R. You can refer to Sections 4.1, 4.2, and 4.3 to try to make your own combination and create some unique applications.

7.1.3.5 R and Hadoop in Real Practice

The technical threshold of combining R and Hadoop is rather high. It requires the programmer not only to master Linux, Java, Hadoop, and R, but also to have the basic knowledge of software development, algorithms, probability statistics, linear algebra, data visualization, and industry background. It demands the coordination of multiple departments and a variety of talents if a company needs to deploy this environment. So cases of combining R with Hadoop in Hadoop operation and maintenance, Hadoop algorithm development, R modeling, MapReduce development in R, and software testing are very rare. This book displays three projects of my attempts and efforts in Sections 6.6, 7.3, and 7.4.

7.1.4 Outlook for the Future

There will be a burst of growth in the combination of R and Hadoop in the near future. But restricted by the technical barrier of interdisciplinary knowledge, there will be fewer talents in this area compared to the market demand. So more and more big data tools will be developed!

7.2 Installation and Use of RHadoop

Question
How do we use R to connect Hadoop?

RHadoop series

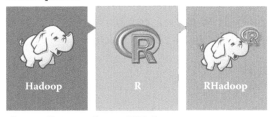

The installation and use of RHadoop
http://blog.fens.me/rhadoop-rhadoop/

RHadoop is the first product to achieve the combination of R and Hadoop and the big data analysis based on it. Hadoop is used for big data storage, and R is used to replace Java to finish the MapReduce algorithms. With the help of RHadoop, R developers now have a powerful tool to process big data at 10G, 100G, TB, or even PB levels. The performance problem of a single machine is perfectly solved. But it's not so easy for R, Java, and Hadoop users to master all the knowledge.

7.2.1 Environment Preparation

For environment preparation, I choose Linux Ubuntu® here. You may choose other types of Linux according to your preferences.

It's important to use the official version Java (JDK) of Oracle SUN and download from the official website (http://www.oracle.com/technetwork/java/javase/downloads/index.html). Linux Ubuntu's own Java (JDK) may have many incompatibility problems. And for JDK, please choose the 1.6.x version, as version 1.7 also has an incompatibility problem. For the R environment, please install version 2.15.3, as version 2.14 and versions above 3.0 do not support RHadoop. System environment in this section is as follows:

- Linux Ubuntu 12.04.2 LTS 64bit server
- R 2.15.3 64bit
- Java JDK 1.6.x
- Hadoop 1.1.2
- IP: 192.168.1.243

Note: RHadoop only supports Linux.

Please refer to Appendix F for information about the installation and configuration of Hadoop.

7.2.2 Installation of RHadoop

RHadoop includes three R packages: rmr, hdfs, and rHBase. Because these three libraries cannot be found in CRAN, we need to download them from other places.

7.2.2.1 Download the Three Relevant Packages of RHadoop

The three packages are rmr-2.1.0, rhdfs-1.0.5, and rHBase-1.1. The address for downloading is https://github.com/RevolutionAnalytics/RHadoop/wiki/Downloads. Though RHadoop has launched rhdfs and rmr2 for Windows version, I still recommend Linux for Hadoop.

7.2.2.2 For the Installation of RHadoop, the Root Authority Operation Is Recommended

```
# Switch to root user.
~ sudo -i
~ whoami
root

# Copy the program package to /root/R.
~ mv  rmr-2.1.0.tar.gz /root/R
~ mv  rhdfs-1.0.5.tar.gz /root/R
~ mv  rhbase-1.1.tar.gz /root/R
```

Then, we need to install the dependant library of these three libraries.

7.2.2.3 Installation of the Dependent Library

Because a RHadoop project depends on Java, we need to install rJava first, which has been explained in Section 4.3.

```
# Configure Java environment variables in R.
~ R CMD javareconf
# Start R.
~ R

# Install rJava.
> install.packages("rJava")
```

Then, we need to install some other dependent libraries directly through install.packages(), including reshape2, Repp, iterators, itertools, digest, RJSONlO, and functional.

```
> install.packages("reshape2")
> install.packages("Rcpp")
> install.packages("iterators")
> install.packages("itertools")
> install.packages("digest")
> install.packages("RJSONIO")
> install.packages("functional")
```

7.2.2.4 Install the rhdfs Library

Use the command export to add HADOOP_CMD and HADOOP_STREAMING to environment variables in the current window. But for the convenience of use, it's good to add the variables to the system environment variable file/etc/environment. Then use R CMD INSTALL command to install rhdfs.

Configure the environment variables in the current environment.

```
~ export HADOOP_CMD=/root/hadoop/hadoop-1.1.2/bin/hadoop
~ export HADOOP_STREAMING=/root/hadoop/hadoop-1.1.2/contrib/
streaming/hadoop-streaming-1.1.2.jar
```

Write the environment variables to/etc/environment.

```
~ sudo vi /etc/environment

HADOOP_CMD=/root/hadoop/hadoop-1.1.2/bin/hadoop
HADOOP_STREAMING=/root/hadoop/hadoop-1.1.2/contrib/streaming/
hadoop-streaming-1.1.2.jar

# Make the environment variables effective.
~ . /etc/environment

# Install rhdfs.
~ R CMD INSTALL /root/R/rhdfs_1.0.5.tar.gz
```

7.2.2.5 Install the rmr Library

Use the R CMD INSTALL command to install rmr library.

```
~ R CMD INSTALL rmr2_2.1.0.tar.gz
```

7.2.2.6 Install the rHBase Library

This is introduced in Section 7.5.

7.2.2.7 List All the Packages of RHadoop

We can check what libraries have been installed in RHadoop.

```
~ ls /disk1/system/usr/local/lib/R/site-library/
digest  functional  iterators  itertools  plyr  Rcpp  reshape2
rhdfs  rJava  RJSONIO  rmr2  stringr
```

Because my hard disk is external and has a mounted directory of R class libraries using mount and symlink(ln –s), my R class libraries are all under the directory/disk1/system. Normal class libraries are in the directory of R is/usr/lib/R/site-library or/usr/local/lib/R/site-library, so you can use them where there is an R command to query the position of R class libraries in your computer.

7.2.3 Program Development of RHadoop

After installing rhdfs and rmr2, we can use R to try some Hadoop operations.

7.2.3.1 Basic Operations of rhdfs

Use of rhdfs.

```
# Load rhdfs.
> library(rhdfs)

Loading required package: rJava
HADOOP_CMD=/root/hadoop/hadoop-1.1.2/bin/hadoop
Be sure to run hdfs.init()

# Initialize R and connect with Hadoop.
> hdfs.init()
```

Compare the Hadoop commands with RHadoop functions.

1. View the file directory of hdfs.
    a. Hadoop command: Hadoop fs –ls.user
    b. R function: hdfs.ls("/user/")

Use the Hadoop command to view the Hadoop directory.

```
~ hadoop fs -ls /user

Found 4 items
drwxr-xr-x   - root supergroup          0 2013-02-01 12:15 /user/conan
drwxr-xr-x   - root supergroup          0 2013-03-06 17:24 /user/hdfs
drwxr-xr-x   - root supergroup          0 2013-02-26 16:51 /user/hive
drwxr-xr-x   - root supergroup          0 2013-03-06 17:21 /user/root
```

Use the rhdfs function to view the Hadoop directory.

```
> hdfs.ls("/user/")

  permission owner     group size            modtime           file
1 drwxr-xr-x  root supergroup    0 2013-02-01 12:15    /user/conan
2 drwxr-xr-x  root supergroup    0 2013-03-06 17:24    /user/hdfs
3 drwxr-xr-x  root supergroup    0 2013-02-26 16:51    /user/hive
4 drwxr-xr-x  root supergroup    0 2013-03-06 17:21    /user/root
```

2. View the data file of Hadoop.
   a. Hadoop command: hadoop fs -cat/user/hdfs/o_same_school/part-m-00000
   b. R function: hdfs.cat("/user/hdfs/o_same_school/part-m-00000")

Use the Hadoop command to view the Hadoop data file.

```
~ hadoop fs -cat /user/hdfs/o_same_school/part-m-00000
10,3,tsinghua university,2004-05-26 15:21:00.0
23,4007,Beijing No.171 Middle School,2004-05-31 06:51:53.0
51,4016,Dalian University of Technology,2004-05-27 09:38:31.0
89,4017,Amherst College,2004-06-01 16:18:56.0
92,4017,Stanford University,2012-11-28 10:33:25.0
99,4017,Stanford University Graduate School of Business,2013-02-19
12:17:15.0
```

Use rhdfs to view the Hadoop data file.

```
> hdfs.cat("/user/hdfs/o_same_school/part-m-00000")
[1] "10,3,tsinghua university,2004-05-26 15:21:00.0"
[2] "23,4007,Beijing No.171 Middle School,2004-05-31 06:51:53.0"
[3] "51,4016,Dalian University of Technology,2004-05-27 09:38:31.0"
[4] "89,4017,Amherst College,2004-06-01 16:18:56.0"
[5] "92,4017,Stanford University,2012-11-28 10:33:25.0"
[6] "99,4017,Stanford University Graduate School of
Business,2013-02-19 12:17:15.0"
```

7.2.3.2 Task of the rmr Algorithm

The following is an introduction to rmr2 and a comparison of normal R and R based on Hadoop.

```
# Load rmr2.
> library(rmr2)

Loading required package: Rcpp
Loading required package: RJSONIO
Loading required package: digest
Loading required package: functional
Loading required package: stringr
Loading required package: plyr
Loading required package: reshape2
```

Normal R:

```
> small.ints = 1:10
> sapply(small.ints, function(x) x^2)

[1]    1    4    9   16   25   36   49   64   81  100
```

R based on Hadoop:

```
> small.ints = to.dfs(1:10)

13/03/07 12:12:55 INFO util.NativeCodeLoader: Loaded the native-
hadoop library
13/03/07 12:12:55 INFO zlib.ZlibFactory: Successfully loaded &
initialized native-zlib library
13/03/07 12:12:55 INFO compress.CodecPool: Got brand-new compressor

# Execute map task.
> mapreduce(input = small.ints, map = function(k, v) cbind(v, v^2))

packageJobJar: [/tmp/RtmpWnzxl4/rmr-local-env5deb2b300d03, /tmp/
RtmpWnzxl4/rmr-global-env5deb398a522b, /tmp/RtmpWnzxl4/rmr-
streaming-map5deb1552172d, /root/hadoop/tmp/hadoop-
unjar7838617732558795635/] [] /tmp/streamjob4380275136001813619.jar
tmpDir=null
13/03/07 12:12:59 INFO mapred.FileInputFormat: Total input paths to
process : 1
13/03/07 12:12:59 INFO streaming.StreamJob: getLocalDirs(): [/root/
hadoop/tmp/mapred/local]
13/03/07 12:12:59 INFO streaming.StreamJob: Running job:
job_201302261738_0293
13/03/07 12:12:59 INFO streaming.StreamJob: To kill this job, run:
13/03/07 12:12:59 INFO streaming.StreamJob: /disk1/hadoop/hadoop-1.1.2/
libexec/../bin/hadoop job -Dmapred.job.tracker=hdfs://192.168.1.243:9001
-kill job_201302261738_0293
```

```
13/03/07 12:12:59 INFO streaming.StreamJob: Tracking URL:
http://192.168.1.243:50030/jobdetails.jsp?jobid=job_201302261738_0293
13/03/07 12:13:00 INFO streaming.StreamJob:  map 0%  reduce 0%
13/03/07 12:13:15 INFO streaming.StreamJob:  map 100%  reduce 0%
13/03/07 12:13:21 INFO streaming.StreamJob:  map 100%  reduce 100%
13/03/07 12:13:21 INFO streaming.StreamJob: Job complete:
job_201302261738_0293
13/03/07 12:13:21 INFO streaming.StreamJob: Output: /tmp/RtmpWnzxl4/
file5deb791fcbd5

# View the output result of hdfs.
> from.dfs("/tmp/RtmpWnzxl4/file5deb791fcbd5")
$key
NULL

$val
       v
 [1,]  1   1
 [2,]  2   4
 [3,]  3   9
 [4,]  4  16
 [5,]  5  25
 [6,]  6  36
 [7,]  7  49
 [8,]  8  64
 [9,]  9  81
[10,] 10 100
```

Because MapReduce can only access the HDFS file system, we need to use to.dfs() to store data toe HDFS system, and then use from.dfs() to extract the computing result of MapReduce from the HDFS file system.

7.2.3.3 Wordcount Task of the rmr Algorithm

Use rmr2 to achieve wordcount task, counting the word in files.

I've placed a data file/user/hdfs/o_same_school/part-m-00000 in HDFS in advance. Execute wordcount(), and use from.dfs() to get the result from HDFS.

```
# Define the position of data file in Hadoop.
> input<- '/user/hdfs/o_same_school/part-m-00000'
# wordcount algorithm function
> wordcount = function(input, output = NULL, pattern = " "){
    # map function
    wc.map = function(., lines) {
            keyval(unlist( strsplit( x = lines,split = pattern)),1)
    }

    # reduce function
    wc.reduce =function(word, counts ) {
            keyval(word, sum(counts))
    }
```

```
    # mapreduce function
    mapreduce(input = input ,output = output, input.format = "text",
        map = wc.map, reduce = wc.reduce,combine = T)
}

# run wordcount task
> wordcount(input)

packageJobJar: [/tmp/RtmpfZUFEa/rmr-local-env6cac64020a8f, /tmp/
RtmpfZUFEa/rmr-global-env6cac73016df3, /tmp/RtmpfZUFEa/rmr-streaming-
map6cac7f145e02, /tmp/RtmpfZUFEa/rmr-streaming-reduce6cac238dbcf, /tmp/
RtmpfZUFEa/rmr-streaming-combine6cac2b9098d4, /root/hadoop/tmp/hadoop-
unjar6584585621285839347/] [] /tmp/streamjob9195921761644130661.jar
tmpDir=null
13/03/07 12:34:41 INFO util.NativeCodeLoader: Loaded the native-hadoop
library
13/03/07 12:34:41 WARN snappy.LoadSnappy: Snappy native library not
loaded
13/03/07 12:34:41 INFO mapred.FileInputFormat: Total input paths to
process : 1
13/03/07 12:34:41 INFO streaming.StreamJob: getLocalDirs(): [/root/
hadoop/tmp/mapred/local]
13/03/07 12:34:41 INFO streaming.StreamJob: Running job:
job_201302261738_0296
13/03/07 12:34:41 INFO streaming.StreamJob: To kill this job, run:
13/03/07 12:34:41 INFO streaming.StreamJob: /disk1/hadoop/hadoop-1.1.2/
libexec/../bin/hadoop job -Dmapred.job.tracker=hdfs://192.168.1.243:9001
-kill job_201302261738_0296
13/03/07 12:34:41 INFO streaming.StreamJob: Tracking URL:
http://192.168.1.243:50030/jobdetails.jsp?jobid=job_201302261738_0296
13/03/07 12:34:42 INFO streaming.StreamJob:  map 0%  reduce 0%
13/03/07 12:34:59 INFO streaming.StreamJob:  map 100%  reduce 0%
13/03/07 12:35:08 INFO streaming.StreamJob:  map 100%  reduce 17%
13/03/07 12:35:14 INFO streaming.StreamJob:  map 100%  reduce 100%
13/03/07 12:35:20 INFO streaming.StreamJob: Job complete:
job_201302261738_0296
13/03/07 12:35:20 INFO streaming.StreamJob: Output: /tmp/RtmpfZUFEa/
file6cac626aa4a7

# View the result of output of hdfs
> from.dfs("/tmp/RtmpfZUFEa/file6cac626aa4a7")

$key
 [1] "-"
 [2] "04:42:37.0"
 [3] "06:51:53.0"
 [4] "07:10:24.0"
 [5] "09:38:31.0"
 [6] "10:33:25.0"
 [7] "10,3,tsinghua"
 [8] "10:42:10.0"
 [9] "113,4017,Stanford"
[10] "12:00:38.0"
```

```
$val
 [1] 1 2 1 2 1 1 1 4 1 1 2 1 1 1 1 2 1 1 2 1 1 1 1 1 1 1 1 1 1 1 1 1
1 1 1 1 1
[39] 1 1 1 1 1 1 1 1 1 1 1 1 1 1 1 1 1 1
```

Thus we've finished the installation and use of rhdfs and rmr2 in RHadoop. Though it's a little tedious, it's still worth it to refine the code of MapReduce in R.

7.3 RHadoop Experiment: Count the Times of the Appearance of Certain E-Mail Addresses

Question
How do we use RHadoop?

**Count the times of the appearance
of certain e-mail addresses
http://blog.fens.me/rhadoop-demo-email/**

We can use R to implement MapReduce development through RHadoop. The R codes are much more concise than the Java codes. This section introduces a case of statistical demand.

7.3.1 Demand Description

System environment used in this section:

- Linux Ubuntu 12.04.2 LTS 64bit server
- R 2.15.3 64bit
- Java JDK 1.6.x
- Hadoop 1.0.3

According to the experimental data, calculate how many times the E-mail addresses appear. Then sort the E-mail addresses by their times of appearance. Use RHadoop to implement the MapReduce algorithm. The experimental data are hadoop15.txt.

```
wolys@21cn.com
zss1984@126.com
294522652@qq.com
simulateboy@163.com
zhoushigang_123@163.com
```

```
sirenxing424@126.com
lixinyu23@qq.com
chenlei1201@gmail.com
370433835@qq.com
cxx0409@126.com
viv093@sina.com
q62148830@163.com
65993266@qq.com
summeredison@sohu.com
zhangbao-autumn@163.com
diduo_007@yahoo.com.cn
fxh852@163.com
weiyang1128@163.com
licaijun007@163.com
junhongshouji@126.com
wuxiaohong11111@163.com
fennal@sina.com
li_dao888@163.com
bokil.xu@163.com
362212053@qq.com
youloveyingying@yahoo.cn
boiny@126.com
linlixian200606@126.com
alex126126@126.com
654468252@qq.com
huangdaqiao@yahoo.com.cn
kitty12502@163.com
xl200811@sohu.com
ysjd8@163.com
851627938@qq.com
wubo_1225@163.com
kangtezc@163.com
xiao2018@126.com
121641873@qq.com
296489419@qq.com
beibeilong012@126.com
```

Output the calculation result as the format below.

```
163.com,14
sohu.com,2
```

7.3.2 Algorithm Implementation

7.3.2.1 Calculate How Many Times the E-Mail Addresses Appear

```
# Start R.
~ R
```

```
> library(rhdfs)
> library(rmr2)

# Load the data to memory from local environment.
> data<-read.table(file="hadoop15.txt")

# Upload the data to hdfs from memory.
> d0<-to.dfs(keyval(1, data))

# View the d0 data on hdfs.
> from.dfs(d0)

$key
[1] 1 1 1 1 1 1 1 1 1 1 1 1 1 1 1 1 1 1 1 1 1 1 1 1 1 1 1 1 1 1 1 1 1 1 1 1 1 1
1 1 1 1 1 1
[39] 1 1 1

$val
V1
1 wolys@21cn.com
2 zss1984@126.com
3 294522652@qq.com
4 simulateboy@163.com
5 zhoushigang_123@163.com
6 sirenxing424@126.com
7 lixinyu23@qq.com
8 chenlei1201@gmail.com
9 370433835@qq.com
10 cxx0409@126.com
11 viv093@sina.com
12 q62148830@163.com
13 65993266@qq.com
14 summeredison@sohu.com
15 zhangbao-autumn@163.com
16 diduo_007@yahoo.com.cn
17 fxh852@163.com
18 weiyang1128@163.com
19 licaijun007@163.com
20 junhongshouji@126.com
21 wuxiaohong11111@163.com
22 fennal@sina.com
23 li_dao888@163.com
24 bokil.xu@163.com
25 362212053@qq.com
26 youloveyingying@yahoo.cn
27 boiny@126.com
28 linlixian200606@126.com
29 alex126126@126.com
30 654468252@qq.com
31 huangdaqiao@yahoo.com.cn
32 kitty12502@163.com
33 xl200811@sohu.com
```

```
34 ysjd8@163.com
35 851627938@qq.com
36 wubo_1225@163.com
37 kangtezc@163.com
38 xiao2018@126.com
39 121641873@qq.com
40 296489419@qq.com
41 beibeilong012@126.com
```

Customize mr() to calculate how many times the E-mail addresses appear.

```
# Create mr().
> mr<-function(input=d0){
+     map<-function(k,v){
+         # Intercept the part of character string after @.
+         keyval(word(as.character(v$V1), 2, sep = fixed('@')),1)
+     }
+     reduce =function(k, v ) {
+         # Sum the same E-mail addresses.
+         keyval(k, sum(v))
+     }
+     d1<-mapreduce(input=input,map=map,reduce=reduce,combine=TRUE)
+ }

# Run mr().
> d1<-mr(d0)

# View the d0 data on hdfs.
> from.dfs(d1)

$key
[1] "126.com" "163.com" "21cn.com" "gmail.com" "qq.com"
[6] "sina.com" "sohu.com" "yahoo.cn" "yahoo.com.cn"

$val
[1] 9 14 1 1 9 2 2 1 2
```

7.3.2.2 Sort the E-Mail Addresses by Their Times of Appearance

Customize sort() to sort the E-mail addresses by their times of appearance.

```
# Create sort().
> sort<-function(input=d1){
+     map<-function(k,v){
+         # Format the result set of d1.
+         keyval(1,data.frame(k,v))
+     }
+     reduce<-function(k,v){
```

```
+          # Rank the data set.
+          v2<-v[order(as.integer(v$v),decreasing=TRUE),]
+          keyval(1,v2)
+       }
+     d2<-mapreduce(input=input,map=map,reduce=reduce,combine=TRUE)
+ }
# Sort.
> d2<-sort(d1)
# Load the result from HDFS to memory.
> result<-from.dfs(d2)

# Output the result of sorting.
> result$val

k v
2 163.com 14
1 126.com 9
5 qq.com 9
6 sina.com 2
7 sohu.com 2
9 yahoo.com.cn 2
3 21cn.com 1
4 gmail.com 1
8 yahoo.cn 1
```

Please note that this section uses statements based on a single node in the reduce process of the second step. It requires more opimization when we deal with big data. We can achieve our objects by just a few lines of code, which is much better than when using Java!

7.4 Implement the Collaborative Filtering Algorithm by RHadoop Based on MapReduce

Question

How do we use RHadoop to implement collaborative filtering algorithm based on MapReduce?

RHadoop series

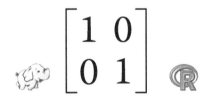

**Collaborative filtering algorithm
by RHadoop based on MapReduce
http://blog.fens.me/rhadoop-mapreduce-rmr/**

Because of the particularity of the operation of RHadoop's rmr2 package on Hadoop, it is rather difficult to implement the codes. If you need to learn it in depth, please try and think more about the design of key/value of the MapReduce algorithm.

7.4.1 Introduction to the Collaborative Filtering Algorithm Based on Item Recommendation

More and more Internet applications now design their own recommendation systems. The recommendation system mainly uses a collaborative filtering algorithm. It can be divided into an algorithm based on users and based on items according to the activity level of users and popularity of items.

- A collaborative filtering algorithm based on users recommends user items appreciated by other users sharing a similar interest.
- A collaborative filtering algorithm based on items recommends user items similar to other items users showed interest in previously.

A collaborative filtering algorithm based on items is more widely used and recommended. Many Internet companies are now using it, including Netflix, YouTube, Amazon, and so forth. There are two main steps in designing the recommendation algorithm:

1. Calculate the similarity between items.
2. Generate a recommendation list according to the similarity of items and historical behavior of users.

To introduce the algorithm model, we select a small group of data as the test set. The data set is taken from the book *Mahout in Action*, p. 49, line 8. The eighth line of original data "3, 101, 2.5" is changed to "3, 101, 2.0". The three fields of each row are user ID, item ID, and rating of item. Create the data file small.csv on server.

```
# Use vi to edit data file.
~ vi small.csv

1,101,5.0
1,102,3.0
1,103,2.5
2,101,2.0
2,102,2.5
2,103,5.0
2,104,2.0
3,101,2.0
3,104,4.0
3,105,4.5
3,107,5.0
4,101,5.0
4,103,3.0
4,104,4.5
4,106,4.0
5,101,4.0
5,102,3.0
5,103,2.0
5,104,4.0
5,105,3.5
5,106,4.0
```

7.4.2 Implementation of a Local Program of R

First, compare the collaborative filtering algorithm based on items by R with that by RHadoop. Here I use the step-by-step algorithm of *Mahout in Action*, Chapter 6. The algorithm is implemented through three steps. First, create a co-occurrence matrix of items; second, create a rating matrix of users on times; and last, calculate the recommendation result through the matrix.

7.4.2.1 Create a Co-Occurrence Matrix of Items

Find the items chosen by each user, and count the times of individual appearance and appearance in pairs. For example, the user of user ID "3" has rated the four items 101, 104, 105, and 107.

1. (101,101), (104,104), (105,105), (107,107), where each individual appearance counts as 1.
2. (101,104), (101,105), (101,107), (104,105), (104,107), (105,107), where each appearance in pairs counts as 1.
3. Sum the results of all users and generate a triangle matrix. Completing the triangle matrix and create the co-occurrence matrix as follows:

```
        [101] [102] [103] [104] [105] [106] [107]
[101]     5     3     4     4     2     2     1
[102]     3     3     3     2     1     1     0
[103]     4     3     4     3     1     2     0
[104]     4     2     3     4     2     2     1
[105]     2     1     1     2     2     1     1
[106]     2     1     2     2     1     2     0
[107]     1     0     0     1     1     0     1
```

7.4.2.2 Create a Rating Matrix of Users on Items

Find all the items and their rating of each user. For example, the user of user ID "3" has rated the four items (3,101,2.0), (3,104,4.0), (3,105,4.5), (3,107,5.0).

1. Find the rating of (3,101,2.0), (3,104,4.0), (3,105,4.5), (3,107,5.0).
2. Create a rating matrix of users on items.

```
        U3
[101]  2.0
[102]  0.0
[103]  0.0
[104]  4.0
[105]  4.5
[106]  0.0
[107]  5.0
```

	101	102	103	104	105	106	107		U3		R
101	5	3	4	4	2	2	1		2.0		40.0
102	3	3	3	2	1	1	0		0.0		18.5
103	4	3	4	3	1	2	0	×	0.0	=	24.5
104	4	2	3	4	2	2	1		4.0		40.0
105	2	1	1	2	2	1	1		4.5		26.0
106	2	1	2	2	1	2	0		0.0		16.5
107	1	0	0	1	1	0	1		5.0		15.5

Figure 7.1   Matrix multiplication. (Excerpted from *Mahout in Action*.)

7.4.2.3 Calculate the Recommendation Result through a Matrix

Co-occurrence * rating matrix = recommendation result, as in Figure 7.1. The result recommended to user "3" is (103,24.5), (102,18.5), (106,16.5).

We implement the aforementioned process using R.

```
# Load plyr.
> library(plyr)
# Load the data set.
> train<-read.csv(file="small.csv",header=FALSE)
> names(train)<-c("user","item","pref")
# View train data set.
> train
  user item pref
1  1  101 5.0
2  1  102 3.0
3  1  103 2.5
4  2  101 2.0
5  2  102 2.5
6  2  103 5.0
7  2  104 2.0
8  3  101 2.0
9  3  104 4.0
10 3  105 4.5
11 3  107 5.0
12 4  101 5.0
13 4  103 3.0
14 4  104 4.5
```

```
15 4 106 4.0
16 5 101 4.0
17 5 102 3.0
18 5 103 2.0
19 5 104 4.0
20 5 105 3.5
21 5 106 4.0
# Method of calculating user list.
> usersUnique<-function(){
+      users<-unique(train$user)
+      users[order(users)]
+ }
# Method of calculating items list.
> itemsUnique<-function(){
+      items<-unique(train$item)
+      items[order(items)]
+ }
# User list.
> users<-usersUnique()
> users
[1] 1 2 3 4 5
# Items list.
> items<-itemsUnique()
> items
[1] 101 102 103 104 105 106 107
# Create item list index.
> index<-function(x) which(items %in% x)
> data<-ddply(train,.(user,item,pref),summarize,idx=index(item))
> data
 user item pref idx
1 1 101 5.0 1
2 1 102 3.0 2
3 1 103 2.5 3
4 2 101 2.0 1
5 2 102 2.5 2
6 2 103 5.0 3
7 2 104 2.0 4
8 3 101 2.0 1
9 3 104 4.0 4
10 3 105 4.5 5
11 3 107 5.0 7
12 4 101 5.0 1
13 4 103 3.0 3
14 4 104 4.5 4
15 4 106 4.0 6
16 5 101 4.0 1
17 5 102 3.0 2
18 5 103 2.0 3
```

```
19 5 104 4.0 4
20 5 105 3.5 5
21 5 106 4.0 6
# Co-occurrence matrix.
> cooccurrence<-function(data){
+       n<-length(items)
+       co<-matrix(rep(0,n*n),nrow=n)
+       for(u in users){
+           idx<-index(data$item[which(data$user==u)])
+           m<-merge(idx,idx)
+           for(i in 1:nrow(m)){
+               co[m$x[i],m$y[i]]=co[m$x[i],m$y[i]]+1
+           }
+       }
+       return(co)
+ }
# Recommendation algorithm.
> recommend<-function(udata=udata,co=coMatrix,num=0){
+       n<-length(items)
+
+       # all of pref
+       pref<-rep(0,n)
+       pref[udata$idx]<-udata$pref
+
+       # User rating matrix.
+       userx<-matrix(pref,nrow=n)
+
+       # Co-occurrence matrix * rating matrix.
+       r<-co %*% userx
+
+       # Rank the recommendation result.
+       r[udata$idx]<-0
+       idx<-order(r,decreasing=TRUE)
+       topn<-data.frame(user=rep(udata$user[1],length(idx)),item=
items[idx],val=r[idx])
+       topn<-topn[which(topn$val>0),]
+
+       # Take the first "num" items of result.
+       if(num>0){
+           topn<-head(topn,num)
+       }
+
+       # Return the result.
+       return(topn)
+ }
# Generate co-occurrence matrix.
> co<-cooccurrence(data)
> co
```

```
      [,1] [,2] [,3] [,4] [,5] [,6] [,7]
[1,]    5    3    4    4    2    2    1
[2,]    3    3    3    2    1    1    0
[3,]    4    3    4    3    1    2    0
[4,]    4    2    3    4    2    2    1
[5,]    2    1    1    2    2    1    1
[6,]    2    1    2    2    1    2    0
[7,]    1    0    0    1    1    0    1
# Calculate recommendation result.
> recommendation<-data.frame()
> for(i in 1:length(users)){
+     udata<-data[which(data$user==users[i]),]
+     recommendation<-rbind(recommendation,recommend(udata,co,0))
+ }
# Output recommendation result.
> recommendation
  user item val
1 1 104 33.5
2 1 106 18.0
3 1 105 15.5
4 1 107 5.0
5 2 106 20.5
6 2 105 15.5
7 2 107 4.0
8 3 103 24.5
9 3 102 18.5
10 3 106 16.5
11 4 102 37.0
12 4 105 26.0
13 4 107 9.5
14 5 107 11.5
```

7.4.3 *Implement a Distributed Program by R Based on Hadoop*

We can use data objects based on R to implement the MapReduce algorithm instead of using file storage as in Java. The pattern of this algorithm is similar to that in R but is more complicated.

1. Create a co-occurrence matrix. First, get the combination list of all items according to user groups; then count the item combination list and create a co-occurrence matrix of items.
2. Create rating matrix of users on items.
3. Merge the co-occurrence matrix and the rating matrix.
4. Calculate the recommendation result list.
5. In MapReduce implementation, all the operations should be completed by using Map and Reduce tasks. The process has changed slightly, as in Figure 7.2. The following are the five steps of matrix analysis.

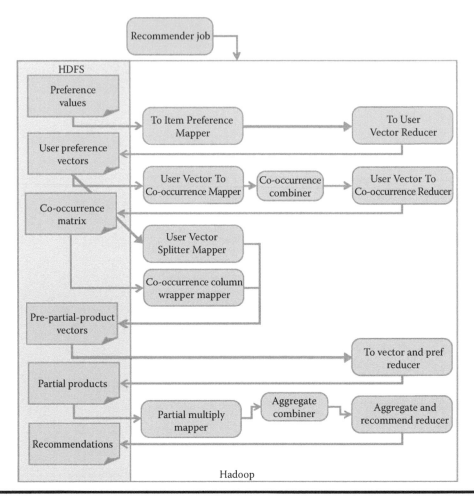

Figure 7.2 Distributed program design. (Excerpted from *Mahout in Action*.)

7.4.3.1 Create a Co-Occurrence Matrix of Items

1. Get the combination list of all items according to user groups.
   a. Key: vector of item list
   b. Val: vector of item combination

```
$key
 [1] 101 101 101 101 101 101 101 101 101 101 101 101 101 101 101 102 102
102 102
[20] 102 102 102 103 103 103 103 103 103 103 103 103 103 103 104 104 104
104 104
[39] 104 104 104 104 104 104 104 105 105 105 105 106 106 106 106 107 107
107 107
[58] 101 101 101 101 101 101 102 102 102 102 102 102 103 103 103 103 103
103 104
[77] 104 104 104 104 104 105 105 105 105 105 105 106 106 106 106 106 106
```

```
$val
[1]  101 102 103 101 102 103 104 101 104 105 107 101 103 104 106 101 102
103 101
[20] 102 103 104 101 102 103 101 102 103 104 101 103 104 106 101 102 103
104 101
[39] 104 105 107 101 103 104 106 101 104 105 107 101 103 104 106 101 104
105 107
[58] 101 102 103 104 105 106 101 102 103 104 105 106 101 102 103 104 105
106 101
[77] 102 103 104 105 106 101 102 103 104 105 106 101 102 103 104 105 106
```

2. Count the item combination list and create co-occurrence matrix of items.
   a. Key: vector of item list
   b. Val: value of data frame of co-occurrence matrix(item, item, Freq)

The format of the matrix should be the same as the format that follows in Section 7.4.3.2. Merge the two heterogeneous data sources into one, laying the foundation for data in Section 7.4.3.3.

```
$key
[1]  101 101 101 101 101 101 101 102 102 102 102 102 102 103 103 103 103
103 103
[20] 104 104 104 104 104 104 104 105 105 105 105 105 105 105 106 106 106
106 106
[39] 106 107 107 107 107

$val
k v freq
1  101 101 5
2  101 102 3
3  101 103 4
4  101 104 4
5  101 105 2
6  101 106 2
7  101 107 1
8  102 101 3
9  102 102 3
10 102 103 3
11 102 104 2
12 102 105 1
13 102 106 1
14 103 101 4
15 103 102 3
16 103 103 4
17 103 104 3
18 103 105 1
19 103 106 2
20 104 101 4
21 104 102 2
22 104 103 3
23 104 104 4
```

```
24  104  105  2
25  104  106  2
26  104  107  1
27  105  101  2
28  105  102  1
29  105  103  1
30  105  104  2
31  105  105  2
32  105  106  1
33  105  107  1
34  106  101  2
35  106  102  1
36  106  103  2
37  106  104  2
38  106  105  1
39  106  106  2
40  107  101  1
41  107  104  1
42  107  105  1
43  107  107  1
```

7.4.3.2 Create a Rating Matrix of Users on Items

- Key: item list
- Val: rating matrix of user on item

```
$key
[1]  101 101 101 101 101 102 102 102 103 103 103 103 104 104 104 104
105 105 106
[20] 106 107

$val
item user pref
1  101 1 5.0
2  101 2 2.0
3  101 3 2.0
4  101 4 5.0
5  101 5 4.0
6  102 1 3.0
7  102 2 2.5
8  102 5 3.0
9  103 1 2.5
10 103 2 5.0
11 103 4 3.0
12 103 5 2.0
13 104 2 2.0
14 104 3 4.0
15 104 4 4.5
16 104 5 4.0
17 105 3 4.5
18 105 5 3.5
```

```
19 106 4 4.0
20 106 5 4.0
21 107 3 5.0
```

7.4.3.3 Merge the Co-Occurrence Matrix and the Rating Matrix

This step is a special operation for MapReduce. We need to process the two heterogeneous data sources for the operation of MapReduce. We've merged the two formats in the second step, and now we use functions in rmr2 to merge matrixes.

- Key: NULL
- Val: merged database

```
$key
NULL

$val
k.1 v.1 freq.1 item.r user.r pref.r
1  103 101 4 103 1 2.5
2  103 102 3 103 1 2.5
3  103 103 4 103 1 2.5
4  103 104 3 103 1 2.5
5  103 105 1 103 1 2.5
6  103 106 2 103 1 2.5
7  103 101 4 103 2 5.0
8  103 102 3 103 2 5.0
9  103 103 4 103 2 5.0
10 103 104 3 103 2 5.0
11 103 105 1 103 2 5.0
12 103 106 2 103 2 5.0
13 103 101 4 103 4 3.0
```

7.4.3.4 Calculate the Recommendation Result List

Calculate the matrix in the third step and get the recommendation result list.

- Key: item list
- Val: data frame of recommendation result

```
$key
[1]  101 101 101 101 101 101 101 101 101 101 101 101 101 101 101 101
101 101
[19] 101 101 101 101 101 101 101 101 101 101 101 101 101 101 101 101
101 102
[37] 102 102 102 102 102 102 102 102 102 102 102 102 102 102 102 102
102 103
[55] 103 103 103 103 103 103 103 103 103 103 103 103 103 103 103 103
103 103
```

```
[73] 103 103 103 103 103 104 104 104 104 104 104 104 104 104 104 104
104 104
[91] 104 104 104 104 104 104 104 104 104 104 104 104 104 104 104 105
105 105
[109] 105 105 105 105 105 105 105 105 105 105 105 106 106 106 106
106 106 106
[127] 106 106 106 106 106 107 107 107 107

$val
k.l v.l user.r v
1 101 101 1 25.0
2 101 101 2 10.0
3 101 101 3 10.0
4 101 101 4 25.0
5 101 101 5 20.0
6 101 102 1 15.0
7 101 102 2 6.0
8 101 102 3 6.0
9 101 102 4 15.0
10 101 102 5 12.0
11 101 103 1 20.0
12 101 103 2 8.0
13 101 103 3 8.0
14 101 103 4 20.0
15 101 103 5 16.0
16 101 104 1 20.0
17 101 104 2 8.0
18 101 104 3 8.0
```

7.4.3.5 Get the Recommendation Rating List in Output Format

Sort the recommendation result list and output the result.

- Key: user ID
- Val: data frame of recommendation result

```
$key
[1] 1 1 1 1 1 1 1 2 2 2 2 2 2 2 3 3 3 3 3 3 3 4 4 4 4 4 4 4 5 5 5 5
5 5 5

$val
user item pref
1 1 101 44.0
2 1 103 39.0
3 1 104 33.5
4 1 102 31.5
5 1 106 18.0
6 1 105 15.5
7 1 107 5.0
8 2 101 45.5
9 2 103 41.5
```

```
10  2  104  36.0
11  2  102  32.5
12  2  106  20.5
13  2  105  15.5
14  2  107  4.0
15  3  101  40.0
16  3  104  38.0
17  3  105  26.0
18  3  103  24.5
19  3  102  18.5
20  3  106  16.5
21  3  107  15.5
22  4  101  63.0
23  4  104  55.0
24  4  103  53.5
25  4  102  37.0
26  4  106  33.0
27  4  105  26.0
28  4  107  9.5
29  5  101  68.0
30  5  104  59.0
31  5  103  56.5
32  5  102  42.5
33  5  106  34.5
34  5  105  32.0
35  5  107  11.5
```

7.4.3.6 Tips for Using rmr2

1. rmr.options(backend = 'hadoop')

   There are two values in "backend" here, "hadoop" and "local." "Hadoop," using a Hadoop environment to run the program, is the default value. "Local" is a configuration for a local test, and it is not recommended now. I've tried "local" in my development. The speed is very fast, and it simulates the running environment of Hadoop. But codes in local mode aren't compatible with Hadoop, and they are often changed, so I don't recommend using "local" here.

2. equijoin(…,outer = c('left'))

   "Outer" here includes four values: c("","left", "right", "full"). It's very much like the "join" operation in database.

3. keyval(k,v)

   We need to store key and valve data in MapReduce operations. If we output the result directly or output the result without adding key, there will be a warning: Converting to.dfs argument to keyval with a NULL key. The example of rmr2 in the last section has a similar circumstance, please note to modify the codes.

```
> to.dfs(1:10)

Warning message:
In to.dfs(1:10) : Converting to.dfs argument to keyval with a NULL key
```

Here is the completed code:

```
# Load rmr2.
> library(rmr2)

# Input data file.
> train<-read.csv(file="small.csv",header=FALSE)
> names(train)<-c("user","item","pref")
# Use the format of Hadoop in rmr. Hadoop is the default configuration.
 > rmr.options(backend = 'hadoop')
# Store the data set in HDFS.
> train.hdfs = to.dfs(keyval(train$user,train))

# Print the result
> from.dfs(train.hdfs)
13/04/07 14:35:44 INFO util.NativeCodeLoader: Loaded the native-hadoop
library
13/04/07 14:35:44 INFO zlib.ZlibFactory: Successfully loaded &
initialized native-zlib library
13/04/07 14:35:44 INFO compress.CodecPool: Got brand-new decompressor

$key
 [1] 1 1 1 2 2 2 2 3 3 3 3 4 4 4 4 5 5 5 5 5 5

$val
   user item pref
1     1  101  5.0
2     1  102  3.0
3     1  103  2.5
4     2  101  2.0
5     2  102  2.5
6     2  103  5.0
7     2  104  2.0
8     3  101  2.0
9     3  104  4.0
10    3  105  4.5
11    3  107  5.0
12    4  101  5.0
13    4  103  3.0
14    4  104  4.5
15    4  106  4.0
16    5  101  4.0
17    5  102  3.0
18    5  103  2.0
19    5  104  4.0
20    5  105  3.5
21    5  106  4.0

# STEP 1, Create co-occurrence matrix of items.
# 1) get the combination list of all items according to user groups.
> train.mr<-mapreduce(
+   train.hdfs,
+   map = function(k, v) {
+     keyval(k,v$item)
+   }
```

```
+    ,reduce=function(k,v){
+       m<-merge(v,v)
+       keyval(m$x,m$y)
+    }
+ )

> from.dfs(train.mr)

$key
 [1]  101 101 101 101 101 101 101 101 101 101 101 101 101 101 101 102
102 102 102
[20]  102 102 102 103 103 103 103 103 103 103 103 103 103 103 104 104
104 104 104
[39]  104 104 104 104 104 104 104 105 105 105 105 106 106 106 106 107
107 107 107
[58]  101 101 101 101 101 101 102 102 102 102 102 102 103 103 103 103
103 103 104
[77]  104 104 104 104 104 105 105 105 105 105 105 106 106 106 106 106 106

$val
 [1]  101 102 103 101 102 103 104 101 104 105 107 101 103 104 106 101
102 103 101
[20]  102 103 104 101 102 103 101 102 103 104 101 103 104 106 101 102
103 104 101
[39]  104 105 107 101 103 104 106 101 104 105 107 101 103 104 106 101
104 105 107
[58]  101 102 103 104 105 106 101 102 103 104 105 106 101 102 103 104
105 106 101
[77]  102 103 104 105 106 101 102 103 104 105 106 101 102 103 104 105 106

# 2) count the item combination list and create co-occurrence matrix
of items.
> step2.mr<-mapreduce(
+    train.mr,
+    map = function(k, v) {
+       d<-data.frame(k,v)
+       d2<-ddply(d,.(k,v),count)
+
+       key<-d2$k
+       val<-d2
+       keyval(key,val)
+    }
+)

> from.dfs(step2.mr)

$key
 [1]  101 101 101 101 101 101 101 102 102 102 102 102 102 103 103 103
103 103 103
[20]  104 104 104 104 104 104 104 105 105 105 105 105 105 105 106 106
106 106 106
[39]  106 107 107 107 107
```

```
$val
     k   v freq
1  101 101    5
2  101 102    3
3  101 103    4
4  101 104    4
5  101 105    2
6  101 106    2
7  101 107    1
8  102 101    3
9  102 102    3
10 102 103    3
11 102 104    2
12 102 105    1
13 102 106    1
14 103 101    4
15 103 102    3
16 103 103    4
17 103 104    3
18 103 105    1
19 103 106    2
20 104 101    4
21 104 102    2
22 104 103    3
23 104 104    4
24 104 105    2
25 104 106    2
26 104 107    1
27 105 101    2
28 105 102    1
29 105 103    1
30 105 104    2
31 105 105    2
32 105 106    1
33 105 107    1
34 106 101    2
35 106 102    1
36 106 103    2
37 106 104    2
38 106 105    1
39 106 106    2
40 107 101    1
41 107 104    1
42 107 105    1
43 107 107    1

# 2. Create rating matrix of users on items.
> train2.mr<-mapreduce(
+   train.hdfs,
+   map = function(k, v) {
+     #df<-v[which(v$user==3),]
+     df<-v
+     key<-df$item
```

```
+     val<-data.frame(item=df$item,user=df$user,pref=df$pref)
+     keyval(key,val)
+   }
+)

> from.dfs(train2.mr)

$key
 [1]  101 101 101 101 101 102 102 102 103 103 103 103 104 104 104 104
105 105 106
[20] 106 107

$val
   item user pref
1   101    1  5.0
2   101    2  2.0
3   101    3  2.0
4   101    4  5.0
5   101    5  4.0
6   102    1  3.0
7   102    2  2.5
8   102    5  3.0
9   103    1  2.5
10  103    2  5.0
11  103    4  3.0
12  103    5  2.0
13  104    2  2.0
14  104    3  4.0
15  104    4  4.5
16  104    5  4.0
17  105    3  4.5
18  105    5  3.5
19  106    4  4.0
20  106    5  4.0
21  107    3  5.0

#3. Merge the co-occurrence matrix and the rating matrix.
> eq.hdfs<-equijoin(
+   left.input=step2.mr,
+   right.input=train2.mr,
+   map.left=function(k,v){
+     keyval(k,v)
+   },
+   map.right=function(k,v){
+     keyval(k,v)
+   },
+   outer = c("left")
+)

> from.dfs(eq.hdfs)

$key
NULL
```

```
$val
    k.l v.l freq.l item.r user.r pref.r
1   103 101      4    103      1    2.5
2   103 102      3    103      1    2.5
3   103 103      4    103      1    2.5
4   103 104      3    103      1    2.5
5   103 105      1    103      1    2.5
6   103 106      2    103      1    2.5
7   103 101      4    103      2    5.0
8   103 102      3    103      2    5.0
9   103 103      4    103      2    5.0
10  103 104      3    103      2    5.0
11  103 105      1    103      2    5.0
12  103 106      2    103      2    5.0
13  103 101      4    103      4    3.0
14  103 102      3    103      4    3.0
15  103 103      4    103      4    3.0
16  103 104      3    103      4    3.0
17  103 105      1    103      4    3.0
18  103 106      2    103      4    3.0
19  103 101      4    103      5    2.0
20  103 102      3    103      5    2.0

# Omit some output.

# 4. Calculate the recommendation result list.
> cal.mr<-mapreduce(
+   input=eq.hdfs,
+   map=function(k,v){
+     val<-v
+     na<-is.na(v$user.r)
+     if(length(which(na))>0) val<-v[-which(is.na(v$user.r)),]
+     keyval(val$k.l,val)
+   }
+   ,reduce=function(k,v){
+     val<-ddply(v,.(k.l,v.l,user.r),summarize,v=freq.l*pref.r)
+     keyval(val$k.l,val)
+   }
+)


> from.dfs(cal.mr)

$key
  [1] 101 101 101 101 101 101 101 101 101 101 101 101 101 101 101
101 101 101
 [19] 101 101 101 101 101 101 101 101 101 101 101 101 101 101 101
101 101 102
 [37] 102 102 102 102 102 102 102 102 102 102 102 102 102 102 102
102 102 103
 [55] 103 103 103 103 103 103 103 103 103 103 103 103 103 103 103
103 103 103
```

```
 [73] 103 103 103 103 103 104 104 104 104 104 104 104 104 104 104
104 104 104
 [91] 104 104 104 104 104 104 104 104 104 104 104 104 104 104 104
105 105 105
[109] 105 105 105 105 105 105 105 105 105 105 105 106 106 106 106
106 106 106
[127] 106 106 106 106 106 107 107 107 107

$val
    k.l v.l user.r     v
1   101 101      1 25.0
2   101 101      2 10.0
3   101 101      3 10.0
4   101 101      4 25.0
5   101 101      5 20.0
6   101 102      1 15.0
7   101 102      2  6.0
8   101 102      3  6.0
9   101 102      4 15.0
10  101 102      5 12.0
11  101 103      1 20.0
12  101 103      2  8.0
13  101 103      3  8.0
14  101 103      4 20.0
15  101 103      5 16.0
16  101 104      1 20.0
17  101 104      2  8.0
18  101 104      3  8.0
19  101 104      4 20.0
20  101 104      5 16.0

# Omit some output.

# 5. Get the recommendation rating list in the format of input.
> result.mr<-mapreduce(
+  input=cal.mr,
+  map=function(k,v){
+    keyval(v$user.r,v)
+  }
+  ,reduce=function(k,v){
+    val<-ddply(v,.(user.r,v.l),summarize,v=sum(v))
+    val2<-val[order(val$v,decreasing=TRUE),]
+    names(val2)<-c("user","item","pref")
+    keyval(val2$user,val2)
+  }
+)

# Print the result
> from.dfs(result.mr)

$key
 [1] 1 1 1 1 1 1 1 2 2 2 2 2 2 2 2 3 3 3 3 3 3 3 4 4 4 4 4 4 5 5 5 5
5 5 5
```

```
$val
   user item pref
1      1  101 44.0
2      1  103 39.0
3      1  104 33.5
4      1  102 31.5
5      1  106 18.0
6      1  105 15.5
7      1  107  5.0
8      2  101 45.5
9      2  103 41.5
10     2  104 36.0
11     2  102 32.5
12     2  106 20.5
13     2  105 15.5
14     2  107  4.0
15     3  101 40.0
16     3  104 38.0
17     3  105 26.0
18     3  103 24.5
19     3  102 18.5
20     3  106 16.5
21     3  107 15.5
22     4  101 63.0
23     4  104 55.0
24     4  103 53.5
25     4  102 37.0
26     4  106 33.0
27     4  105 26.0
28     4  107  9.5
29     5  101 68.0
30     5  104 59.0
31     5  103 56.5
32     5  102 42.5
33     5  106 34.5
34     5  105 32.0
35     5  107 11.5
```

This section provides a method of using R to implement a collaborative filtering algorithm based on MapReduce. The algorithm might not be optimal, so I hope you can write a better one! With the development of R and Hadoop, I believe that there will be more and more algorithms using this method.

If you want to compare this with the collaborative filtering algorithm based on MapReduce of Java, please refer to the author's blog "*Use Hadoop to Construct File Recommendation System*" and "*Step-by-Step Program Development of Mahout: Collaborative Filtering ItemCF Based on Items.*"

7.5 Installation and Use of RHBase

Question

How do we use R to access HBase?

RHadoop series

Installation and use of RHBase
http://blog.fens.me/rhadoop-hbase-rhase/

HBase is a distributed database product of the Hadoop family. It has many advantages, including supporting high-concurrency reading and writing, column-style data storage, efficient index, auto-sharing, and auto region migration. It has become more and more accepted and used in industry.

7.5.1 Environment Preparation of HBase

Here we continue to use the environment in Section 7.2.

- Linux Ubuntu 12.04.2 LTS 64bit server
- Java JDK 1.6.0_45
- Hadoop 1.1.2
- HBase-0.94.2
- thrift-0.8.0

For information on the installation and configuration of HBase, please refer to Appendix H. View the server environment of HBase. Use command bin/start-hbase.sh to start HBase server. The default port is port = 60000.

```
# Start Hadoop.
~ /home/conan/hadoop/hadoop-1.1.2/bin/start-all.sh
# Start HBase.
~ /home/conan/hadoop/hbase-0.94.2/bin/start-hbase.sh
# Start Thrift service of HBase.
~ /home/conan/hadoop/hbase-0.94.2/bin/hbase-daemon.sh start thrift
# View HBase process.
~ jps
12041 HMaster
12209 HRegionServer
13222 ThriftServer
31734 TaskTracker
31343 DataNode
31499 SecondaryNameNode
13328 Jps
31596 JobTracker
11916 HQuorumPeer
31216 NameNode
```

7.5.2 Installation of rHBase

We downloaded rHBase-1.1.tar.gz in Section 7.2, so I'll skip the introduction to rHBase. We need only one line of command to finish the installation of rHBase.

```
~ R CMD INSTALL rhbase_1.1.1.tar.gz
```

7.5.3 Function Library of rHBase

RHBase supports 16 functions to implement operations of HBase.

7.5.3.1 16 Functions of rHBase

The following are the 16 functions and their comparison with those in HBase.

```
hb.compact.table   hb.describe.table   hb.insert            hb.regions.table
hb.defaults        hb.get              hb.insert.data.frame    hb.scan

hb.delete          hb.get.data.frame   hb.list.tables       hb.scan.ex
hb.delete.table    hb.init             hb.new.table         hb.set.table.mode
```

7.5.3.2 Basic Operations between HBase and rHBase

```
# Create table.
HBase Shell:
create 'student_shell','info'
rhbase:
hb.new.table("student_rhbase","info")
# List all the tables.
HBase Shell:
list
rhbase:
hb.list.tables()
# Display table structure.
HBase Shell:
describe 'student_shell'
rhbase:
hb.describe.table("student_rhbase")
# Insert a piece of data.
HBase Shell:
put 'student_shell','mary','info:age','19'
rhbase:
hb.insert("student_rhbase",list(list("mary","info:age", "24")))
# Load data.
HBase Shell:
get 'student_shell','mary'
```

```
rhbase:
hb.get('student_rhbase','mary')
# Delete table(We need two commands to delete table in HBase, and
only one in rHBase).
HBase Shell:
disable 'student_shell'
drop 'student_shell'
rhbase:
hb.delete.table('student_rhbase')
```

7.5.3.3 Execute Base Operations of HBase

We use HBase Shell and rHBase to execute basic function operations on HBase. First is HBase Shell.

```
# Start HBase Shell.
~ /home/conan/hadoop/hbase-0.94.18/bin/hbase shell
# Create table.
> create 'student_shell','info'
> list
TABLE
student_shell
# Display table structure.
> describe 'student_shell'
DESCRIPTION
ENABLED
{NAME => 'student_shell', FAMILIES => [{NAME => 'info', DATA_BLOCK_
true
ENCODING => 'NONE', BLOOMFILTER => 'NONE', REPLICATION_SCOPE => '0'
, VERSIONS => '3', COMPRESSION => 'NONE', MIN_VERSIONS => '0', TTL
=> '2147483647', KEEP_DELETED_CELLS => 'false', BLOCKSIZE => '65536
', IN_MEMORY => 'false', ENCODE_ON_DISK => 'true', BLOCKCACHE =>
'true'}]}
# Insert a piece of data.
>  put 'student_shell','mary','info:age','19'
# View data.
>  get 'student_shell','mary'
COLUMN                        CELL
info:age                      timestamp=1365414964962, value=19
# Delete table.
> disable 'student_shell'
> drop 'student_shell'
```

Then use rHBase to perform the same operations.

```
# Start R.
~ R
```

```
# Load rHBase.
> library(rhbase)
# Initialize HBase in R.
> hb.init()
<pointer: 0x16494a0>
attr(,"class")
[1] "hb.client.connection"

# Create table.
> hb.new.table("student_rhbase","info",opts=list(maxversions=5,x=
list(maxversions=1L,compression='GZ',inmemory=TRUE)))
[1] TRUE
# View all the tables.
> hb.list.tables()
$student_rhbase
  maxversions compression inmemory bloomfiltertype bloomfiltervecsize
info:           5        NONE    FALSE            NONE
0
      bloomfilternbhashes blockcache timetolive
info:                 0      FALSE        -1
# View table structure.
> hb.describe.table("student_rhbase")
      maxversions compression inmemory bloomfiltertype
bloomfiltervecsize
info:           5        NONE    FALSE            NONE
0
      bloomfilternbhashes blockcache timetolive
info:                 0      FALSE        -1

# Insert one piece of data.
> hb.insert("student_rhbase",list(list("mary","info:age", "24")))
[1] TRUE

# Query data.
> hb.get('student_rhbase','mary')
[[1]]
[[1]][[1]]
[1] "mary"

[[1]][[2]]
[1] "info:age"

[[1]][[3]]
[[1]][[3]][[1]]
[1] "24"

# Delete table.
> hb.delete.table('student_rhbase')
[1] TRUE
```

Now we've accessed HBase using R.

7.6 Solve the Installation Error of RHadoop PipeMapRed.waitOutputThreads()

Question

In the installation process of RHadoop, we may meet the error PipeMapRed.waitOutputThreads(). How do we solve the error?

RHadoop series

Solve the installation error of RHadoop

PipeMapRed.waitOutputThreads () : subprocess failed with code 1

http://blog.fens.me/rhadoo-rmr2-pipemapred/

When we use the rmr2 package in RHadoop, we may meet the error PipeMapRed.waitOutput Threads(): subprocess failed with code 1. This error may have troubled many users. What's the reason for this error? And how do we solve it? This section gives you the answers.

The code for possible error is as follows.

```
> small.ints = to.dfs(1:10)
> mapreduce(input = small.ints, map = function(k, v) cbind(v, v^2))
> from.dfs("/tmp/RtmpWnzxl4/file5deb791fcbd5")
```

7.6.1 Error Log of rmr2

System environment in this section:

- Linux Ubuntu 12.04.2 LTS 64bit server
- R 2.15.3 64bit
- Java JDK 1.6.x
- Hadoop 1.1.2
- IP: 192.168.1.243

Error log in R:

```
packageJobJar: [/tmp/Rtmpdf7egm/rmr-local-env3cbb70983, /tmp/
Rtmpdf7egm/rmr-global-env3cbb654b85fe,/tmp/Rtmpdf7egm/rmr-streaming-
map3cbba213f2e, /home/conan/hadoop/tmp/]
```

```
hadoop-unjar1697638502297829404/] [] /tmp/
streamjob4620072667602885650.jar tmpDir=null
13/06/23 10:44:25 INFO mapred.FileInputFormat: Total input paths to
process : 1
13/06/23 10:44:25 INFO streaming.StreamJob: getLocalDirs(): [/home/
conan/hadoop/tmp/mapred/local]
13/06/23 10:44:25 INFO streaming.StreamJob: Running job:
job_201306231032_0001
13/06/23 10:44:25 INFO streaming.StreamJob: To kill this job, run:
13/06/23 10:44:25 INFO streaming.StreamJob: /home/conan/hadoop/
hadoop-1.1.2/libexec/../bin/hadoop job  -Dmapred.job.tracker=hdfs://
master:9001 -kill job_201306231032_0001
13/06/23 10:44:25 INFO streaming.StreamJob: Tracking URL: http://
master:50030/jobdetails.jsp?jobid=job_201306231032_0001
13/06/23 10:44:26 INFO streaming.StreamJob:  map 0%  reduce 0%
13/06/23 10:45:04 INFO streaming.StreamJob:  map 100%  reduce 100%
13/06/23 10:45:04 INFO streaming.StreamJob: To kill this job, run:
13/06/23 10:45:04 INFO streaming.StreamJob: /home/conan/hadoop/
hadoop-1.1.2/libexec/../bin/hadoop job  -Dmapred.job.tracker=hdfs://
master:9001 -kill job_201306231032_0001
13/06/23 10:45:04 INFO streaming.StreamJob: Tracking URL: http://
master:50030/jobdetails.jsp?jobid=job_201306231032_0001
13/06/23 10:45:04 ERROR streaming.StreamJob: Job not successful.
Error: # of failed Map Tasks exceeded allowed limit. FailedCount: 1.
LastFailedTask: task_201306231032_0001_m_000000
13/06/23 10:45:04 INFO streaming.StreamJob: killJob...
Streaming Command Failed!
Error in mr(map = map, reduce = reduce, combine = combine,
vectorized.reduce, :
  hadoop streaming failed with error code 1
```

We can't find the actual error of Hadoop only from the log above.

7.6.2 Locate the Error in the Hadoop Log

Then we need to locate the position of the error. To query the log conveniently, open the console jobtracker (http://192.168.1.210:50030/jobtracker.jsp) and find the webpage location of the log of error, as in Figure 7.3.

View the error of map. Now it has a clear definition.

```
Running a job using hadoop streaming and mrjob: PipeMapRed.
waitOutputThreads(): subprocess failed with code 1
java.lang.RuntimeException: PipeMapRed.waitOutputThreads():
subprocess failed with code 1 at
org.apache.hadoop.streaming.PipeMapRed.waitOutputThreads(PipeMapRed.
java:362) at
```

```
org.apache.hadoop.streaming.PipeMapRed.mapRedFinished(PipeMapRed.
java:576) at
org.apache.hadoop.streaming.PipeMapper.close(PipeMapper.java:135) at
org.apache.hadoop.mapred.MapRunner.run(MapRunner.java:57) at
org.apache.hadoop.streaming.PipeMapRunner.run(PipeMapRunner.java:36) at
org.apache.hadoop.mapred.MapTask.runOldMapper(MapTask.java:436) at
org.apache.hadoop.mapred.MapTask.run(MapTask.java:372) at
org.apache.hadoop.mapred.Child$4.run(Child.java:255) at
java.security.AccessController.doPrivileged(Native Method) at
javax.security.auth.Subject.doAs(Subject.java:396) at
org.apache.hadoop.security.UserGroupInformation.
doAs(UserGroupInformation.java:1121) at
org.apache.hadoop.mapred.Child.main(Child.java:249)
```

Then we use two methods to troubleshoot and solve the error.

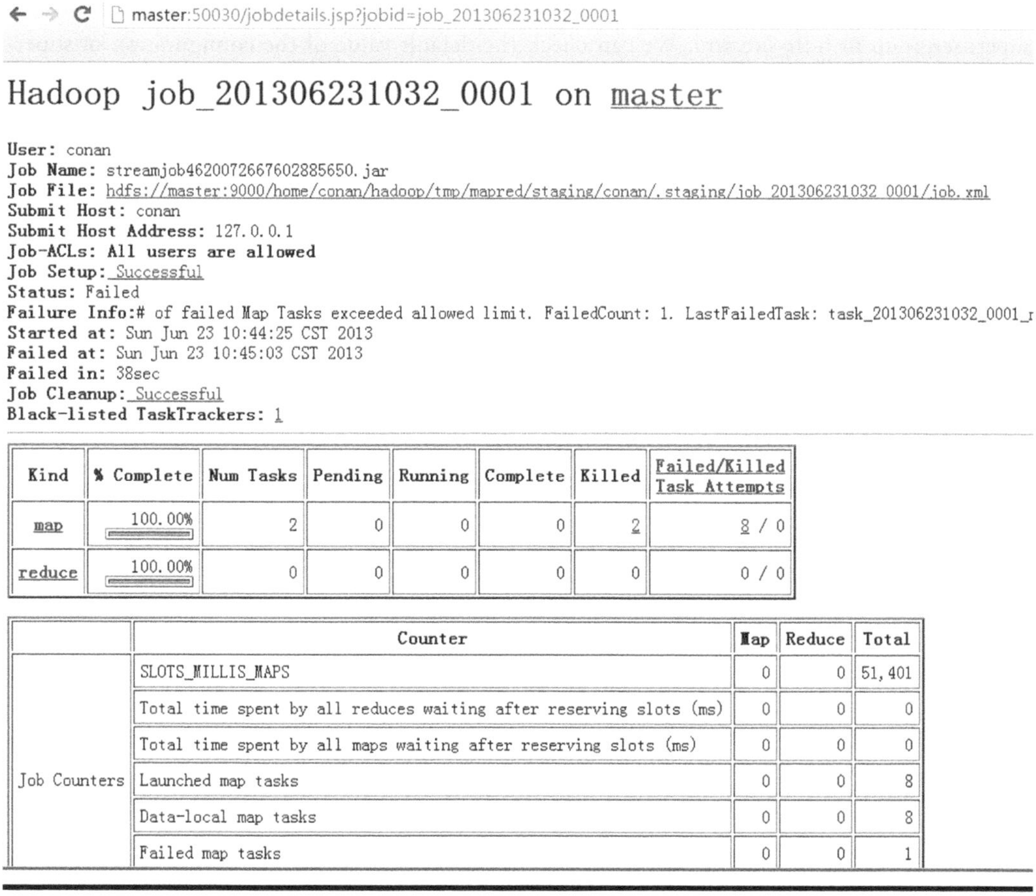
master:50030/jobdetails.jsp?jobid=job_201306231032_0001

Hadoop job_201306231032_0001 on master

User: conan
Job Name: streamjob4620072667602885650.jar
Job File: hdfs://master:9000/home/conan/hadoop/tmp/mapred/staging/conan/.staging/job_201306231032_0001/job.xml
Submit Host: conan
Submit Host Address: 127.0.0.1
Job-ACLs: All users are allowed
Job Setup: Successful
Status: Failed
Failure Info:# of failed Map Tasks exceeded allowed limit. FailedCount: 1. LastFailedTask: task_201306231032_0001_r
Started at: Sun Jun 23 10:44:25 CST 2013
Failed at: Sun Jun 23 10:45:03 CST 2013
Failed in: 38sec
Job Cleanup: Successful
Black-listed TaskTrackers: 1

Kind	% Complete	Num Tasks	Pending	Running	Complete	Killed	Failed/Killed Task Attempts
map	100.00%	2	0	0	0	2	8 / 0
reduce	100.00%	0	0	0	0	0	0 / 0

	Counter	Map	Reduce	Total
Job Counters	SLOTS_MILLIS_MAPS	0	0	51,401
	Total time spent by all reduces waiting after reserving slots (ms)	0	0	0
	Total time spent by all maps waiting after reserving slots (ms)	0	0	0
	Launched map tasks	0	0	8
	Data-local map tasks	0	0	8
	Failed map tasks	0	0	1

Figure 7.3 Console of Jobtracker.

7.6.3 Find Solutions in Hadoop: Failed

View the error list of Hadoop: error of code 1.

```
"OS error code 1: Operation not permitted""OS error code 2: No such
file or directory"
```

From its description, we may find that this is an error of authority. Let's make an analysis of the authority of users and user groups in RHadoop environment.

■ Initiating user of Hadoop: user: conan, user group: conan
■ Initiating user of R: user: conan, user group: conan

The users have the same authority, so why would an error appear? I've run another test in Hadoop using the root authority, where no error is reported.

Now because the problem is quite specific, let's look for the reason for this error on the Internet. Search "org.apache.hadoop.security.AccessControlException" on Google. According to the search result on the Internet, adjust the configuration of Hadoop: add the definition of dfs.permissions .superusergroup in hdfs-site.xml. We can check the default value of the configuration of super-usergroup in the system.

The default name of the group is now supergourp. Modify hdfs-site.xml and add the definition of dfs.permissions.superusergroup.

```
~ vi $HADOOP_HOME/conf/hdfs-site.xml

<configuration>
    <property>
        <name>dfs.data.dir</name>
        <value>/home/conan/hadoop/data</value>
    </property>
    <property>
        <name>dfs.replication</name>
        <value>1</value>
    </property>
    <property>
        <name>dfs.permissions</name>
        <value>false</value>
    </property>
    <property>
        <name>dfs.permissions.superusergroup</name>
        <value>supergroup</value>
    </property>
</configuration>
```

Restart the Hadoop cluster and R, and run rmr2 script. But the error is still not solved.

7.6.4 Find Solutions in RHadoop: Succeeded

Looking at the issue 122 of RHadoop on Github (https://github.com/RevolutionAnalytics /RHadoop/issues/122), we find that others also face similar problems. But the issue is closed now, indicating that this problem has been solved according to the answer, "The problem was indeed that the rmr package (and other dependent packages) needed to be installed in the system directory rather than the user specific directory." In my own environment, the configuration is the same as the preceding description.

```
# Start R.
~ R

# View the installation directory of R packages.
> .libPaths()
[1] "/home/conan/R/x86_64-pc-linux-gnu-library/2.15"
[2] "/usr/local/lib/R/site-library"
[3] "/usr/lib/R/site-library"
[4] "/usr/lib/R/library"

# View the R packages using commands.
~ ls /home/conan/R/x86_64-pc-linux-gnu-library/2.15
colorspace  functional  iterators  munsell  RColorBrewer  rhdfs
rmr2
dichromat   ggplot2     itertools  plyr     Rcpp          rJava
scales
digest      gtable      labeling   proto    reshape2      RJSONIO
stringr

# /usr/local is empty.
~ ls /usr/local/lib/R/site-library
```

I find that all the class libraries in my environment are under the directory/home/conan. And there is none under the directory/usr/local/lib/R/site-library. So we reinstall the RHadoop using root authority according to the direction of the author of RHadoop.

```
# Switch to root.
~ sudo -i
# View the path of class libraries of R.
~ R
> .libPaths()
[1] "/usr/local/lib/R/site-library" "/usr/lib/R/site-library"
[3] "/usr/lib/R/library"
# Install under root authority.
~ R CMD javareconf
~ R
```

```
# Start R.
> install.packages("rJava")
> install.packages("reshape2")
> install.packages("Rcpp")
> install.packages("iterators")
> install.packages("itertools")
> install.packages("digest")
> install.packages("RJSONIO")
> install.packages("functional")

~ cd /home/conan/R
~ R CMD INSTALL rmr2_2.1.0.tar.gz

# There are R packages under /usr/local.
~ ls /usr/local/lib/R/site-library
digest  functional  iterators  itertools  plyr  Rcpp  reshape2
rJava  RJSONIO  rmr2  stringr
```

The dependent packages are installed under/usr/local/lib/R/site-library. Quit root user and restart R to test.

```
# Quit root user.
~ exit
~ whoami
conan

# Restart R.
~ R
> library(rmr2)
Loading required package: Rcpp
Loading required package: RJSONIO
Loading required package: digest
Loading required package: functional
Loading required package: stringr
Loading required package: plyr
Loading required package: reshape

# Run rmr2.
> small.ints = to.dfs(1:10)
> mapreduce(input = small.ints, map = function(k, v) cbind(v, v^2))

packageJobJar: [/tmp/RtmpM87JEc/rmr-local-env1c7588ca7ed, /tmp/
RtmpM87JEc/rmr-global-env1c77fdcab5f, /tmp/RtmpM87JEc/rmr-streaming-
map1c76a4ddf6e, /home/conan/hadoop/tmp/hadoop-
unjar6992113986427459004/] [] /tmp/streamjob2762947354578034435.jar
tmpDir=null
13/06/23 13:27:36 INFO mapred.FileInputFormat: Total input paths to
process : 1
13/06/23 13:27:36 INFO streaming.StreamJob: getLocalDirs(): [/home/
conan/hadoop/tmp/mapred/local]
```

```
13/06/23 13:27:36 INFO streaming.StreamJob: Running job:
job_201306231141_0007
13/06/23 13:27:36 INFO streaming.StreamJob: To kill this job, run:
13/06/23 13:27:36 INFO streaming.StreamJob: /home/conan/hadoop/
hadoop-1.1.2/libexec/../bin/hadoop job  -Dmapred.job.tracker=hdfs://
master:9001 -kill job_201306231141_0007
13/06/23 13:27:36 INFO streaming.StreamJob: Tracking URL: http://
master:50030/jobdetails.jsp?jobid=job_201306231141_0007
13/06/23 13:27:37 INFO streaming.StreamJob:  map 0%  reduce 0%
13/06/23 13:27:51 INFO streaming.StreamJob:  map 100%  reduce 0%
13/06/23 13:27:58 INFO streaming.StreamJob:  map 100%  reduce 100%
13/06/23 13:27:58 INFO streaming.StreamJob: Job complete:
job_201306231141_0007
13/06/23 13:27:58 INFO streaming.StreamJob: Output: /tmp/RtmpM87JEc/
file1c722c5c6ae
```

The problem is finally solved!

This section has solved the aforementioned problem in four steps: finding the error, locating the error, finding the reason, and solving the error. I hope that this section will not only help you with this problem, but also provide you with some ideas to improve your ability to solve other problems on your own!

The body of this book is now finished. Thanks for reading. I hope that this book may open a new door for further study. The next book of this series, *R for Geeks: Advanced Development*, will introduce more content for learning R. I hope that readers can develop some interesting applications using the tools and techniques introduced in this book.

APPENDIXES

Appendix A: Installation of Java Environment

A.1 Installation of Java in Windows®

Installation of Java environment refers to the installation of a Java development environment, including the installation of JDK and environment configuration of Java. Many Java applications can be run through JRE in Windows. But Java developers should install JDK officially published by Oracle SUN.

It's very simple to install JDK in Windows by downloading and running the exe file, available from http://www.oracle.com/technetwork/java/javase/downloads/index.html. We can download the latest version Java SE 8, or the earlier JDK versions. JDK version 1.6.0_45 is used in this book, installed in the directory D:\toolkit\java\jdk6. After installation, we should start the environment variables configuration, as in Figure A.1. Select Control Panel → System and Security → System Attribute → Advanced → Environment Variables.

```
# Create JAVA_HOME variable.
JAVA_HOME=D:\toolkit\java\jdk6

# Create PATH Variable.
PATH=D:\toolkit\java\jdk6\bin
```

Note: If the PATH variable already exists, we should add the path of JDK to the last and separate it from the last configuration with a semicolon(;).

After saving the result, open a new CMD window and test the command line operation of Java.

```
# Check the version of Java.
~ C:\Users\Administrator>java -version
java version "1.6.0_45"
Java(TM) SE Runtime Environment (build 1.6.0_45-b06)
Java HotSpot(TM) 64-Bit Server VM (build 20.45-b01, mixed mode)
```

```
# Test Java command.
~ C:\Users\Administrator>java
Usage: java [-options] class [args...]
           (to execute a class)
   or  java [-options] -jar jarfile [args...]
           (to execute a jar file)
where options include:
    -server         to select the "server" VM
    -hotspot        is a synonym for the "server" VM  [deprecated]
                    The default VM is server.
    -cp <class search path of directories and zip/jar files>
    -classpath <class search path of directories and zip/jar files>
                    A ; separated list of directories, JAR archives,
                    and ZIP archives to search for class files.
    -D<name>=<value>
                    set a system property
    -verbose[:class|gc|jni]
                    enable verbose output
    -version        print product version and exit
    -version:<value>
                    require the specified version to run
    -showversion    print product version and continue
    -jre-restrict-search | -jre-no-restrict-search
                    include/exclude user private JREs in the version
search
    -? -help        print this help message
    -X              print help on non-standard options
    -ea[:<packagename>...|:<classname>]
    -enableassertions[:<packagename>...|:<classname>]
                    enable assertions
    -da[:<packagename>...|:<classname>]
    -disableassertions[:<packagename>...|:<classname>]
                    disable assertions
    -esa | -enablesystemassertions
                    enable system assertions
    -dsa | -disablesystemassertions
                    disable system assertions
    -agentlib:<libname>[=<options>]
                    load native agent library <libname>, e.g.
-agentlib:hprof
                    see also, -agentlib:jdwp=help and
-agentlib:hprof=help
    -agentpath:<pathname>[=<options>]
                    load native agent library by full pathname
    -javaagent:<jarpath>[=<options>]
                    load Java programming language agent, see java.
lang.instrument
    -splash:<imagepath>
                    show splash screen with specified image

# Test Javac command.
~ C:\Users\Administrator>javac
```

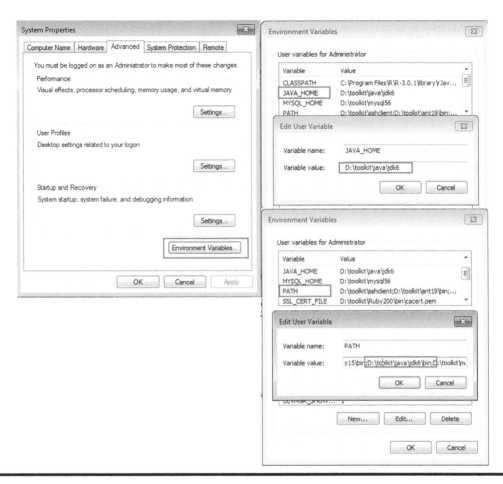

Figure A.1   Configuration of Java environment variables.

Thus the Java environment configuration of Java in Windows is completed.

A.2 Installation of Java in Linux Ubuntu®

The installation of Java here refers to installing the official JDK software packages of Oracle SUN, not installing OpenJDK through apt-get in Linux Ubuntu. I choose to install the 64-bit version Java, Java SE 1.6.45, in Linux Ubuntu. Please pay attention to the version number and use the official JDK software package of Oracle SUN.

A.2.1 Download JDK

Now we need to sign into the official website of Oracle to download JDK. We cannot download JDK by directly using the wget command. Here are two solutions: (1) Sign in and download JDK in your browser, and then upload it to the Linux server. (2) Use the wget command to perform simulate signing in through user cookies, and then download. A further explanation of the second solution is available on the author's blog, http://blog.fens.me/linux-java-install.

A.2.2 Install JDK

Install JDK through the Shell command.

```
# Install JDK.
~ sh ./jdk-6u45-linux-x64.bin
~ pwd
/home/conan/tookit

# Check the directory after installation.
~ tree -L 2
.
├── jdk1.6.0_45
│       ├── bin
│       ├── COPYRIGHT
│       ├── db
│       ├── include
│       ├── jre
│       ├── lib
│       ├── LICENSE
│       ├── man
│       ├── README.html
│       ├── src.zip
│       └── THIRDPARTYLICENSEREADME.txt
└── jdk-6u45-linux-x64.bin
```

A.2.3 Set the Environment Variables of Java

Edit the environment variable file/etc/environment using vi.

```
~ sudo vi /etc/environment

PATH="/usr/local/sbin:/usr/local/bin:/usr/sbin:/usr/bin:/sbin:/bin:/
usr/games:/home/conan/tookit/jdk1.6.0_45/bin"
JAVA_HOME=/home/conan/tookit/jdk1.6.0_45
```

Check the configuration of environment variables.

```
# Take the environment variables into effect.
~ . /etc/environment

# Check the environment variables.
~ export|grep jdk
declare -x OLDPWD="/home/conan/tookit/jdk1.6.0_45"
declare -x
```

```
PATH="/usr/local/sbin:/usr/local/bin:/usr/sbin:/usr/bin:/sbin:/bin:/
usr/games:/home/conan/tookit/jdk1.6.0_45/bin"

~ echo $JAVA_HOME
/home/conan/tookit/jdk1.6.0_45
```

Run the Java command.

```
~ java -version
java version "1.6.0_45"
Java(TM) SE Runtime Environment (build 1.6.0_45-b06)
Java HotSpot(TM) 64-Bit Server VM (build 20.45-b01, mixed mode)
~ java
Usage: java [-options] class [args...]
          (to execute a class)
   or  java [-options] -jar jarfile [args...]
          (to execute a jar file)
where options include:
    -d32          use a 32-bit data model if available
    -d64          use a 64-bit data model if available
    -server       to select the "server" VM
                  The default VM is server.

    -cp
    -classpath
                  A : separated list of directories, JAR archives,
                  and ZIP archives to search for class files.
    -D=
                  set a system property
    -verbose[:class|gc|jni]
                  enable verbose output
    -version      print product version and exit
    -version:
                  require the specified version to run
    -showversion  print product version and continue
    -jre-restrict-search | -jre-no-restrict-search
                  include/exclude user private JREs in the version
                  search
    -? -help      print this help message
    -X            print help on non-standard options
    -ea[:...|:]
    -enableassertions[:...|:]
                  enable assertions
    -da[:...|:]
    -disableassertions[:...|:]
                  disable assertions
    -esa | -enablesystemassertions
                  enable system assertions
    -dsa | -disablesystemassertions
                  disable system assertions
```

```
-agentlib:[=]
               load native agent library, e.g. -agentlib:hprof
                 see also, -agentlib:jdwp=help and
                 -agentlib:hprof=help
-agentpath:[=]
               load native agent library by full pathname
-javaagent:[=]
               load Java programming language agent, see java.
               lang.instrument
-splash:
               show splash screen with specified image
```

After all of the preceding operations, the installation of a Java environment in Linux Ubuntu is completed.

Appendix B: Installation of MySQL

B.1 Installation of MySQL in Windows®

It's very simple to install MySQL in Windows by downloading and decompressing the package. The address is http://dev.mysql.com/downloads/mysql/.

- Command for Running of MySQL server: installation directory of MySQL/bin/mysqld.exe
- Command for Running of MySQL client: installation directory of MySQL/bin/mysql.exe

B.2 Installation of MySQL in Linux Ubuntu®

This book uses Linux Ubuntu 12.04.2 LTS 64bit. The installation of MySQL package can be implemented through apt-get. First install MySQL in Linux Ubuntu.

```
# Install MySQL server.
~ sudo apt-get install mysql-server
```

During the process we need to input the password of the root user. I choose mysql as my password. After installation, the MySQL server will start automatically and we can check the program of the MySQL server.

```
# Check the system process of MySQL server.
~ ps -aux|grep mysql
mysql    3205  2.0  0.5 549896 44092 ?      Ssl  20:10  0:00 /usr/
sbin/mysqld
conan    3360  0.0  0.0  11064   928 pts/0 S+   20:10  0:00 grep
--color=auto mysql

# Check the port used in the MySQL server.
~ netstat -nlt|grep 3306
tcp      0      0 127.0.0.1:3306              0.0.0.0:*
LISTEN
```

```
# Check the state of MySQL server through start command.
~ sudo /etc/init.d/mysql status
Rather than invoking init scripts through /etc/init.d, use the
service(8)
utility, e.g. service mysql status
Since the script you are attempting to invoke has been converted to an
Upstart job, you may also use the status(8) utility, e.g. status mysql
mysql start/running, process 3205

# Check the sate of MySQL server through system service command.
~ service mysql status
mysql start/running, process 3205
```

B.3 Access MySQL through the Command Line Client

The client program of the MySQL command line will be installed automatically with MySQL server. Input mysql command to start the client program to access the MySQL server.

```
~ mysql
Welcome to the MySQL monitor.  Commands end with ; or \g.
Your MySQL connection id is 42
Server version: 5.5.35-0ubuntu0.12.04.2 (Ubuntu)

Copyright (c) 2000, 2013, Oracle and/or its affiliates. All rights
reserved.

Oracle is a registered trademark of Oracle Corporation and/or its
affiliates. Other names may be trademarks of their respective
owners.

Type 'help;' or '\h' for help. Type '\c' to clear the current input
statement.

mysql>
```

Input the username and password to login to the server.

```
~ mysql -uroot -p
Enter password:
Welcome to the MySQL monitor.  Commands end with ; or \g.
Your MySQL connection id is 37
Server version: 5.5.35-0ubuntu0.12.04.2 (Ubuntu)

Copyright (c) 2000, 2013, Oracle and/or its affiliates. All rights
reserved.
```

```
Oracle is a registered trademark of Oracle Corporation and/or its
affiliates. Other names may be trademarks of their respective owners.

Type 'help;' or '\h' for help. Type '\c' to clear the current input
statement.

mysql>
```

The following is a simple command operation of MySQL.

```
# View all the databases.
mysql> show databases;
+--------------------+
| Database           |
+--------------------+
| information_schema |
| test               |
+--------------------+
# View the character set encoding of database.
mysql> show variables like '%char%';
+--------------------------+----------------------------+
| Variable_name            | Value                      |
+--------------------------+----------------------------+
| character_set_client     | utf8                       |
| character_set_connection | utf8                       |
| character_set_database   | utf8                       |
| character_set_filesystem | binary                     |
| character_set_results    | utf8                       |
| character_set_server     | latin1                     |
| character_set_system     | utf8                       |
| character_sets_dir       | /usr/share/mysql/charsets/ |
+--------------------------+----------------------------+
8 rows in set (0.00 sec)
```

B.4 Modify the Configuration of MySQL Server

We need to modify the configuration to make MySQL adapted to the demand of development.

B.4.1 Set the Character Encoding to UTF-8

As the default character set of MySQL is latin1, Chinese characters would be turned to garbled words when they are stored. So we need to change the character set to UTF-8.

Use vi to open the configuration file, my.cnf, of MySQL server.

```
~ sudo vi /etc/mysql/my.cnf

# Add the character encoding of client under the label [client].
[client]
default-character-set=utf8
```

```
# Add the character encoding of client under the label [mysqld].
[mysqld]
character-set-server=utf8
collation-server=utf8_general_ci
```

B.4.2 Permit MySQL Server to Be Accessed Remotely

By default, the MySQL server can be accessed only locally but not remotely. So we need to change the option of remote access. Open the configuration file of MySQL server, my.cnf, using vi.

```
~ sudo vi /etc/mysql/my.cnf

#bind-address            = 127.0.0.1
```

Restart the MySQL server after modification.

```
~ sudo /etc/init.d/mysql restart
Rather than invoking init scripts through /etc/init.d, use the
service(8)
utility, e.g. service mysql restart

Since the script you are attempting to invoke has been converted to an
Upstart job, you may also use the stop(8) and then start(8)
utilities,
e.g. stop mysql ; start mysql. The restart(8) utility is also
available.
mysql start/running, process 3577
```

Re-login to the server.

```
~ mysql -uroot -p
Enter password:
Welcome to the MySQL monitor.  Commands end with ; or \g.
Your MySQL connection id is 37
Server version: 5.5.35-0ubuntu0.12.04.2 (Ubuntu)

Copyright (c) 2000, 2013, Oracle and/or its affiliates. All rights
reserved.

Oracle is a registered trademark of Oracle Corporation and/or its
affiliates. Other names may be trademarks of their respective
owners.

Type 'help;' or '\h' for help. Type '\c' to clear the current input
statement.
```

```
# View the character set encoding again.
mysql> show variables like '%char%';
+--------------------------+----------------------------+
| Variable_name            | Value                      |
+--------------------------+----------------------------+
| character_set_client     | utf8                       |
| character_set_connection | utf8                       |
| character_set_database   | utf8                       |
| character_set_filesystem | binary                     |
| character_set_results    | utf8                       |
| character_set_server     | utf8                       |
| character_set_system     | utf8                       |
| character_sets_dir       | /usr/share/mysql/charsets/ |
+--------------------------+----------------------------+
8 rows in set (0.00 sec)
```

Check the network monitoring port of MySQL.

```
# Check the port used in MySQL server.
~ netstat -nlt|grep 3306
  tcp        0      0 0.0.0.0:3306            0.0.0.0:*
LISTEN
```

We can see that the network monitoring is transformed to 0 0.0.0.0.0:3306 from 127.0.0.1:3306, which indicates that MySQL can be accessed remotely now.

Now the MySQL server is installed in Linux Ubuntu.

Appendix C: Installation of Redis

C.1 Installation of Redis in Windows®

It's very simple to install Redis in Windows by downloading and running the exe file available from https://github.com/rgl/redis/downloads.

- Command for running of Redis server: installation directory of Redis/redis-server.exe
- Command for running of Redis client: installation directory of Redis/redis-cli.exe

C.2 Installation of Redis in Linux Ubuntu®

This book uses Linux Ubuntu 12.04.2 LTS 64bit. The installation of Redis packages can be achieved through apt-get.

```
# Install Redis server.
~ sudo apt-get install redis-server
```

After the installation, Redis server will start automatically. Now let's check the program of the Redis server.

```
# Check the system process of Redis server.
~ ps -aux|grep redis
redis     4162  0.1  0.0  10676  1420 ?        Ss    23:24   0:00 /
usr/bin/redis-server /etc/redis/redis.conf
conan     4172  0.0  0.0  11064   924 pts/0    S+    23:26   0:00
grep --color=auto redis

# Check the condition of Redis server through start command.
~ netstat -nlt|grep 6379
tcp       0       0 127.0.0.1:6379           0.0.0.0:*
LISTEN
```

```
# Check the condition of Redis server through start command.
~ sudo /etc/init.d/redis-server status
redis-server is running
```

C.3 Access Redis through the Command Line Client

The program of the command line client of Redis will be installed automatically with the Redis server. Input the command redis-cli and the client program will access the Redis server.

```
# Start Redis command line client.
~ redis-cli
redis 127.0.0.1:6379>

# Help of command line.
redis 127.0.0.1:6379> help
redis-cli 2.2.12
Type: "help @" to get a list of commands in
      "help " for help on
      "help " to get a list of possible help topics
      "quit" to exit

# View the table of all keys.
redis 127.0.0.1:6379> keys *
(empty list or set)
```

C.4 Modify the Configuration of Redis

C.4.1 Use the Access Account of Redis

By default, we don't need login authentication to access Redis server. To increase the security level, we can set an access password of the Redis server, redisredis as in this case.

Open the configuration file of Redis server, redis.conf, using vi.

```
~ sudo vi /etc/redis/redis.conf

# Cancel the comment "requirepass". The access password is redisredis.
```

C.4.2 Permit Redis Server to Be Accessed Remotely

By default, the Redis server can be accessed only locally but not remotely. So we need to change the option of remote access. Open the configuration file of Redis server, my.cnf, using vi.

```
~ sudo vi /etc/redis/redis.conf

#bind 127.0.0.1
```

Restart the Redis server after modification.

```
~ sudo /etc/init.d/redis-server restart
Stopping redis-server: redis-server.
Starting redis-server: redis-server.
```

Sign into the Redis server without using a password.

```
~ redis-cli

redis 127.0.0.1:6379> keys *
(error) ERR operation not permitted
```

We can sign into the server but can't run a command. We then sign into the Redis server and input a password.

```
~  redis-cli -a redisredis

redis 127.0.0.1:6379> keys *
1) "key2"
2) "key3"
3) "key4"
```

Everything goes well this time. Check the network monitoring port of Redis.

```
# Check the port used in Redis server.
~ netstat -nlt|grep 6379
tcp        0      0 0.0.0.0:6379            0.0.0.0:*               LISTEN
```

We can see that the network monitoring is transformed to 0 0.0.0.0.0:3306 from 127.0.0.1:3306, which indicates that Redis can be accessed remotely now. We can access the Redis server remotely from another Linux.

```
~ redis-cli -a redisredis -h 192.168.1.199

redis 192.168.1.199:6379> keys *
1) "key2"
2) "key3"
3) "key4"
```

Remote access is successful. Now the Redis server is installed in Linux Ubuntu.

Appendix D: Installation of MongoDB

D.1 Installation of MongoDB in Windows®

It's very easy to install MongoDB in Windows: download the executable file and install it directly. The address for downloadinbg is http://www.mongodb.org/downloads.

- Running command of MongoDB server: installation directory/bin/mongod.exe
- Running command of MongoDB client: installation directory/bin/mongo.exe

D.2 Installation of MongoDB in Linux Ubuntu®

This book uses Linux Ubuntu 12.04.2 LTS 64bit. We can install MongoDB using apt-get, but we need to download the software source of MongoDB from the official website, https://www.mongodb.org /downloads. First, we need to modify the file source.list of apt-get and add the configuration of 10gen.

```
# Download key file.
~ sudo apt-key adv --keyserver hkp://keyserver.ubuntu.com:80 --recv 7F0CEB10
Executing: gpg --ignore-time-conflict --no-options --no-default-
keyring --secret-keyring /tmp/tmp.kVFab9XYw0 --trustdb-name /etc/apt/
trustdb.gpg --keyring /etc/apt/trusted.gpg --primary-keyring /etc/apt/
trusted.gpg --keyserver hkp://keyserver.ubuntu.com:80 --recv 7F0CEB10
gpg: Download key '7F0CEB10' from hkp server keyserver.ubuntu.com
gpg: Key 7F0CEB10: public key "Richard Kreuter" has been imported.
gpg: No key of absolute trust is found.
gpg: The number of processed: 1
gpg:                    Number of imported: 1   (RSA: 1)

# Add the configuration of software source of MongoDB in source.list.
~ echo 'deb http://downloads-distro.mongodb.org/repo/ubuntu-upstart
dist 10gen' | sudo tee /etc/apt/sources.list.d/mongodb.list
deb http://downloads-distro.mongodb.org/repo/ubuntu-upstart dist 10gen

# Update the list of software source.
~ sudo apt-get update
```

Install MongoDB in Linux Ubuntu.

```
# Install MongoDB server.
~ sudo apt-get install mongodb-10gen
```

After installation, MongoDB will be started automatically. Let's check the program of the MongoDB server.

```
# Check the system process of MongoDB server.
~  ps -aux|grep mongo
mongodb  6870  3.7  0.4 349208 39740 ?        Ssl  10:27   2:23 /
usr/bin/mongod --config /etc/mongodb.conf

# Check the status of MongoDB server through start-up command.
~  netstat -nlt|grep 27017
tcp       0       0 0.0.0.0:27017           0.0.0.0:*            LISTEN

# Check the status of MongoDB server through start-up command.
~ sudo /etc/init.d/mongodb status
Rather than invoking init scripts through /etc/init.d, use the service(8)
utility, e.g. service mongodb status

Since the script you are attempting to invoke has been converted to an
Upstart job, you may also use the status(8) utility, e.g. status mongodb
mongodb start/running, process 6870

# Check the status of MongoDB server through system service.
~ sudo service mongodb status
mongodb start/running, process 6870
```

Check the status of the MongoDB server through the console of Web. Enter http://ip:28017 in the browser and open the console of the Web, as in Figure D.1.

D.3 Access MongoDB through the Command Line Client

The client program of the MongoDB command line will be installed automatically with the MongoDB server. Enter the command mongo and the client program will be started to access the MongoDB server.

```
# Start MongoDB client.
~ mongo
MongoDB shell version: 2.4.9
connecting to: test
Welcome to the MongoDB shell.
For interactive help, type "help".
For more comprehensive documentation, see

http://docs.mongodb.org/

Questions? Try the support group

http://groups.google.com/group/mongodb-user

# View the help information of command line.
> help

db.help()                       help on db methods
db.mycoll.help()                help on collection methods
sh.help()                       sharding helpers
rs.help()                       replica set helpers
help admin                      administrative help
help connect                    connecting to a db help
help keys                       key shortcuts
help misc                       misc things to know
help mr                         mapreduce

show dbs                        show database names
show collections                show collections in current database
show users                      show users in current database
show profile                    show most recent system.profile entries
with time >= 1ms
show logs                       show the accessible logger names
show log [name]                 prints out the last segment of log in
memory, 'global' is default
use                 set current database
db.foo.find()                   list objects in collection foo
db.foo.find( { a : 1 } )        list objects in foo where a == 1
it                              result of the last line evaluated; use
to further iterate
DBQuery.shellBatchSize = x    set default number of items to display
on shell
exit                            quit the mongo shell
```

The MongoDB server allows external access by default. Such single-node MongoDB is successfully installed in Linux Ubuntu.

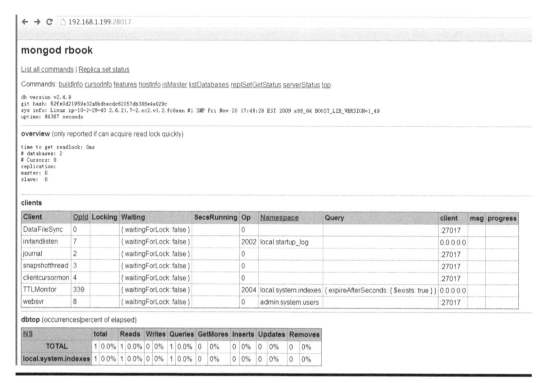

Figure D.1   Web console of MongoDB.

Appendix E: Installation of Cassandra

E.1 Environment Preparation of Ubuntu®

Cassandra, based on Java, is an NoSQL software. It doesn't provide a Windows® version, so I'll just introduce the installation of Cassandra in Linux Ubuntu. As Cassandra is developed by Java, we need to install a Java environment first (please refer to Appendix A).

Cassandra doesn't provide the software source of apt-get for installation, so we need to download the Cassandra package on the official website, http://cassandra.apache.org/download/. This section uses version 1.2.15 as an example and performs the installation and configuration. Please find the corresponding version and download it.

System environment:

- Linux Ubuntu 12.04.2 LTS 64bit server
- Java JDK 1.6.0_45

E.2 Download the Cassandra Package

Download Cassandra:

```
# Download the version 1.2.15.
~ wget http://apache.dataguru.cn/cassandra/1.2.15/apache
-cassandra-1.2.15-bin.tar.gz

# Decompress the package.
~ tar xvf apache-cassandra-1.2.15-bin.tar.gz

# View the file and the directory.
~ tree -L 2
.
```

```
├── apache-cassandra-1.2.15
│   ├── bin
│   ├── CHANGES.txt
│   ├── conf
│   ├── interface
│   ├── javadoc
│   ├── lib
│   ├── LICENSE.txt
│   ├── NEWS.txt
│   ├── NOTICE.txt
│   ├── pylib
│   ├── README.txt
│   └── tools
├── apache-cassandra-1.2.15-bin.tar.gz

# Change the name of the decompressed directory.
~ mv apache-cassandra-1.2.15/ cassandra1215

# Enter the directory.
~ cd cassandra1215/
```

E.3 Configure Cassandra

Configure the data directory of Cassandra.

- Data_file_directories: data file directory
- Commitlog_directory: log file directory
- Saved_caches_directory: cache file directory

Use vi open the configuration file cassandra.yaml of Cassandra.

```
~ vi conf/cassandra.yaml

data_file_directories:
    - /var/lib/cassandra/data
commitlog_directory: /var/lib/cassandra/commitlog
saved_caches_directory: /var/lib/cassandra/saved_caches
```

Make sure that all these directories have been created in the operating system and the directory /var/log/Cassandra is writable by users.

```
# Create directory.
~ sudo mkdir -p /var/lib/cassandra/data
~ sudo mkdir -p /var/lib/cassandra/saved_caches
~ sudo mkdir -p /var/lib/cassandra/commitlog
~ sudo mkdir -p /var/log/cassandra/
```

```
# Assign the directories to users.
~ sudo chown -R conan:conan /var/lib/cassandra
~ sudo chown -R conan:conan /var/log/cassandra/

# View the directory authority.
~ ll /var/lib/cassandra
drwxr-xr-x  2 conan conan 4096  3月 22 06:23 commitlog/
drwxr-xr-x  2 conan conan 4096  3月 22 06:23 data/
drwxr-xr-x  2 conan conan 4096  3月 22 06:23 saved_caches/
```

E.4 Set the Environment Variables

```
~ sudo vi /etc/environment
CASSANDRA_HOME=/home/conan/toolkit/cassandra1215

# Take the environment variables into effect.
~ . /etc/environment

# View the environment variables.
~ echo $CASSANDRA_HOME
/home/conan/toolkit/cassandra1215
```

E.5 Start the Cassandra Server

Start the Cassandra server through commands.

```
# The parameter -f is to bind to console. If there is no -f, then it
will be started in the background.
~ bin/cassandra

# View the system process of Cassandra.
~ ps -axu|grep cassandra

# View the system port.
~ netstat -nlt|grep 9160
tcp        0      0 127.0.0.1:9160          0.0.0.0:*           LISTEN
```

E.6 Use Client to Access Cassandra

Use the client program to access Cassandra server.

```
~ bin/cassandra-cli
Connected to: "Test Cluster" on 127.0.0.1/9160
Welcome to Cassandra CLI version 1.2.15
```

```
Type 'help;' or '?' for help.
Type 'quit;' or 'exit;' to quit.

# View the help information of command line.
[default@unknown] ?
```

Now single-node Cassandra is installed in the Linux Ubuntu system.

Appendix F: Installation of Hadoop

F.1 Environment Preparation of Ubuntu

Hadoop supports the development on Linux®, Windows®, and Mac, but it cannot be deployed in Windows and Mac on a large scale, so I'll focus on the installation and use of Hadoop in Linux Ubuntu.

Because Hadoop is developed by Java, we need to install a Java environment first. For the installation of Java, please refer to Appendix A. Hadoop doesn't provide an apt-get software source for installation, so we need to download the Hadoop package from the official website, http://hadoop.apache.org/#Download+Hadoop.

System environment:

- Linux Ubuntu 12.04.2 LTS 64bit server
- Java JDK 1.6.0_45

F.2 Find the Historical Version of Hadoop

Here we need version 1.1.2 of Hadoop. Open the page for downloading Hadoop, http://www.apache.org/dyn/closer.cgi/hadoop/common/. We can't find version 1.1.2 in those download mirrors, as in Figure F.1.

Then check the distribution source of Hadoop on Github, as in Figure F.2. Find release 1.1.2.

```
https://github.com/apache/hadoop-common/releases/tag/release-1.1.2
```

Use the Linux command wget to download the source code.

```
~ wget https://github.com/apache/hadoop-common/archive/release-
1.1.2.tar.gz
```

← → C | 🗋 apache.dataguru.cn/hadoop/common/   ☆

Index of /hadoop/common/

```
../
alpha/                          16-Aug-2013 05:26        -
beta/                           15-Aug-2013 21:10        -
hadoop-0.23.9/                  01-Jul-2013 17:16        -
hadoop-1.2.1/                   22-Jul-2013 22:49        -
hadoop-2.0.3-alpha/             07-Feb-2013 03:53        -
hadoop-2.0.4-alpha/             12-Apr-2013 21:53        -
hadoop-2.0.5-alpha/             04-Jun-2013 21:48        -
hadoop-2.0.6-alpha/             16-Aug-2013 05:26        -
hadoop-2.1.0-beta/             15-Aug-2013 21:10        -
stable/                         22-Jul-2013 22:49        -
HEADER.html                     23-Jan-2008 21:44      419
readme.txt                      13-Dec-2012 05:19      187
```

Figure F.1 Hadoop mirror.

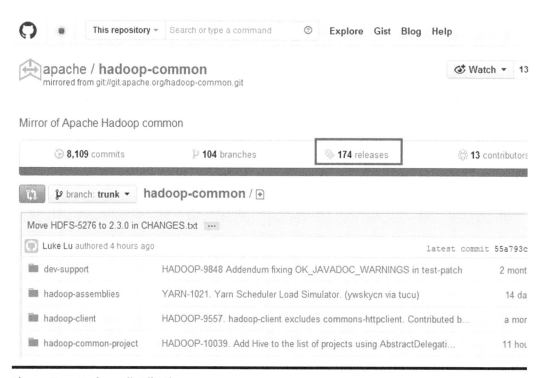

Figure F.2 Hadoop distribution source.

Check branch-1.1 on Github.

```
https://github.com/apache/hadoop-common/tree/branch-1.1
```

F.3 Use the Source Code to Construct the Hadoop Environment

Linux system environment:

■ Linux Ubuntu 64bit Server 12.04.2 LTS
■ Java 1.6.0_29
■ Ant 1.8.4

Note: The structure of latest hadoop-3.0.0-SNAPSHOT on Github is completely changed. Use Maven to replace the construction process of Ant + Ivy.

F.3.1 Install Hadoop

```
~ tar xvf release-1.1.2.tar.gz
~ mkdir/home/conan/hadoop/
~ mv hadoop-common-release-1.1.2/home/conan/hadoop/
~ cd/home/conan/hadoop/
~ mv hadoop-common-release-1.1.2/hadoop-1.1.2
```

View the Hadoop directory.

```
~ cd  hadoop-1.1.2

~ ls -l
total 648
drwxrwxr-x  2 conan conan   4096 Mar  6  2013 bin
-rw-rw-r--  1 conan conan 120025 Mar  6  2013 build.xml
-rw-rw-r--  1 conan conan 467130 Mar  6  2013 CHANGES.txt
drwxrwxr-x  2 conan conan   4096 Oct  3 02:31 conf
drwxrwxr-x  2 conan conan   4096 Oct  3 02:28 ivy
-rw-rw-r--  1 conan conan  10525 Mar  6  2013 ivy.xml
drwxrwxr-x  4 conan conan   4096 Mar  6  2013 lib
-rw-rw-r--  1 conan conan  13366 Mar  6  2013 LICENSE.txt
drwxrwxr-x  2 conan conan   4096 Oct  3 03:35 logs
-rw-rw-r--  1 conan conan    101 Mar  6  2013 NOTICE.txt
-rw-rw-r--  1 conan conan   1366 Mar  6  2013 README.txt
-rw-rw-r--  1 conan conan   7815 Mar  6  2013 sample-conf.tgz
drwxrwxr-x 16 conan conan   4096 Mar  6  2013 src
```

There is no all kinds of class libraries of Hadoop-*.jar or dependent library under the root directory.

F.3.2 Use Ant to Compile

Download Ant from the official website (http://ant.apache.org/bindownload.cgi).

```
~ wget http://archive.apache.org/dist/ant/binaries/apache-ant-1.8.4-
bin.tar.gz
~ tar xvf apache-ant-1.8.4-bin.tar.gz
~ mkdir/home/conan/toolkit/
~ mv apache-ant-1.8.4/home/conan/toolkit/
~ cd/home/conan/toolkit/
~ mv apache-ant-1.8.4 ant184
```

Set Ant to environment variable.

```
# Edit environment file.
~  sudo vi /etc/environment

PATH="/usr/local/sbin:/usr/local/bin:/usr/sbin:/usr/bin:/sbin:/bin:/
usr/games:/home/conan/toolkit/jdk16/bin:/home/conan/toolkit/ant184/bin"

JAVA_HOME=/home/conan/toolkit/jdk16
ANT_HOME=/home/conan/toolkit/ant184

# Take the environment variable into effect.
~ . /etc/environment

# Run Ant command to check whether the environment variable is configured.
~ ant
Buildfile: build.xml does not exist!
Build failed
```

Use Ant to compile Hadoop.

```
# Install the library for compiling.
~ sudo apt-get install autoconf
~ sudo apt-get install libtool

# Use Ant to compile Hadoop.
~ cd /home/conan/hadoop/hadoop-1.1.2/
~ ant
```

We should wait several minutes to build successfully. Then check the build directory generated.

```
~ ls -l build
drwxrwxwr-x  3 conan conan    4096 Oct   3 04:06 ant
drwxrwxwr-x  2 conan conan    4096 Oct   3 04:02 c++
```

```
drwxrwxr-x  3 conan conan    4096 Oct  3 04:05 classes
drwxrwxr-x 13 conan conan    4096 Oct  3 04:06 contrib
drwxrwxr-x  2 conan conan    4096 Oct  3 04:05 empty
drwxrwxr-x  2 conan conan    4096 Oct  3 04:02 examples
-rw-rw-r--  1 conan conan     423 Oct  3 04:05 hadoop-client-1.1.3-
SNAPSHOT.jar
-rw-rw-r--  1 conan conan 4035744 Oct  3 04:05 hadoop-core-1.1.3-
SNAPSHOT.jar
-rw-rw-r--  1 conan conan     426 Oct  3 04:05 hadoop-minicluster-
1.1.3-SNAPSHOT.jar
-rw-rw-r--  1 conan conan  306827 Oct  3 04:05 hadoop-tools-1.1.3-
SNAPSHOT.jar
drwxrwxr-x  4 conan conan    4096 Oct  3 04:02 ivy
drwxrwxr-x  3 conan conan    4096 Oct  3 04:02 src
drwxrwxr-x  6 conan conan    4096 Oct  3 04:02 test
drwxrwxr-x  3 conan conan    4096 Oct  3 04:05 tools
drwxrwxr-x  9 conan conan    4096 Oct  3 04:02 webapps
```

I find that Hadoop-.jar all end with hadoop—1.1.3-SNAPSHOT.jar. The possible explanation is that the release of the last version is the SNAPSHOT of the next version. Modify the 31st line of build.xml and make the value of attribute version 1.1.2. Thus the name of packages generated will be hadoop-*-1.1.2.jar. Restart Ant.

```
~ vi build.xml

<property name = "version" value = "1.1.2"/>

~ rm -rf build
~ ant
```

Now we've installed the Hadoop environment.

F.4 Configure the Environment Script of Hadoop Quickly

Perform the configuration following the next seven steps: configure environment variables, set the three configuration files of Hadoop, create the Hadoop directory, configure the hostname and hosts, generate the SSH login-free key, format HDFS, and start the Hadoop service.

1. Configure the system environment variables.

```
~ sudo vi /etc/environment

PATH="/usr/local/sbin:/usr/local/bin:/usr/sbin:/usr/bin:/sbin:/bin:/
usr/games:/home/conan/toolkit/jdk16/bin:/home/conan/toolkit/ant184/
bin:/home/conan/toolkit/maven3/bin:/home/conan/toolkit/tomcat7/bin:/
home/conan/hadoop/hadoop-1.1.2/bin"
```

```
JAVA_HOME=/home/conan/toolkit/jdk16
ANT_HOME=/home/conan/toolkit/ant184
MAVEN_HOME=/home/conan/toolkit/maven3
HADOOP_HOME=/home/conan/hadoop/hadoop-1.1.2
HADOOP_CMD=/home/conan/hadoop/hadoop-1.1.2/bin/hadoop
HADOOP_STREAMING=/home/conan/hadoop/hadoop-1.1.2/contrib/streaming/
hadoop-streaming-1.1.2.jar

# Take the environment variables into effect.
~ . /etc/environment
```

2. Set the three configuration files of Hadoop.

```
#core-site.xml
~ vi conf/core-site.xml
<configuration>
<property>
<name>fs.default.name</name>
<value>hdfs://master:9000</value>
</property>
<property>
<name>hadoop.tmp.dir</name>
<value>/home/conan/hadoop/tmp</value>
</property>
<property>
<name>io.sort.mb</name>
<value>256</value>
</property>
</configuration>

#hdfs-site.xml
~ vi conf/hdfs-site.xml
<configuration>
<property>
<name>dfs.data.dir</name>
<value>/home/conan/hadoop/data</value>
</property>
<property>
<name>dfs.replication</name>
<value>1</value>
</property>
<property>
<name>dfs.permissions</name>
<value>false</value>
</property>
</configuration>

#mapred-site.xml
~ vi conf/mapred-site.xml
<configuration>
```

```
<property>
<name>mapred.job.tracker</name>
<value>hdfs://master:9001</value>
</property>
</configuration>
```

3. Create a Hadoop directory.

```
~ mkdir /home/conan/hadoop/data
~ mkdir /home/conan/hadoop/tmp
~ sudo chmod 755 /home/conan/hadoop/data/
~ sudo chmod 755 /home/conan/hadoop/tmp/
```

4. Configure the hostname and hosts.

```
~ sudo hostname master
~ sudo vi /etc/hosts
192.168.1.210    master
127.0.0.1        localhost
```

5. Generate an SSH login-free key.

```
~ ssh-keygen -t rsa
~ cat ~/.ssh/id_rsa.pub >>  ~/.ssh/authorized_keys
```

6. Format HDFS.

```
~ bin/hadoop namenode -format
```

7. Start the Hadoop service.

```
~ bin/start-all.sh
```

Check the running status of Hadoop.

```
~ jps
15574 DataNode
16324 Jps
15858 SecondaryNameNode
```

```
16241 TaskTracker
15283 NameNode
15942 JobTracker
```

Check the running status of Hadoop nodes.

```
~ bin/hadoop dfsadmin -report
```

F.5 Compile hadoop-core.jar for a Windows Environment

In a Windows environment, we would often meet an exception indicating a wrong authority check. We need to modify line 688 to line 693 of comments in FileUtil.java.

```
~ vi src/core/org/apache/hadoop/fs/FileUtil.java

685 private static void checkReturnValue(boolean rv, File p,
686                                      FsPermission permission
687                                      ) throws IOException {
688    /*  if (!rv) {
689      throw new IOException("Failed to set permissions of path: " + p +
690                            " to " +
691                            String.format("%04o", permission.
toShort()));
692      }
693    */
694    }
```

After using Ant to repack, hadoop-core-1.1.2.jar that can be run in Windows is generated! We've installed single-node Hadoop in Linux Ubuntu.

Appendix G: Installation of the Hive Environment

G.1 Installation of Hive

After installing the Hadoop environment, we can install Hive on namenode. For the environment configuration of Hadoop, please refer to Appendix F. For the installation of Hive, first download hive-0.9.0.tar.gz, then decompress to/home/cos/toolkit/hive-0.9.0, enter hive directory, and create configuration file.

```
~ cd/home/cos/toolkit/hive-0.9.0

# Copy the configuration file.
~ cp hive-default.xml.template hive-site.xml

# Copy the log file.
~ cp hive-log4j.properties.template hive-log4j.properties
```

Modify the configuration file hive-site.xml to sotreo metadata of Hive to MySQL.

```
~ vi conf/hive-site.xml

<property>
<name>javax.jdo.option.ConnectionURL</name>
<value>jdbc:mysql://localhost:3306/hive_metadata?createDatabaseIfNot
Exist =
true</value>
<description>JDBC connect string for a JDBC metastore</description>
</property>

<property>
<name>javax.jdo.option.ConnectionDriverName</name>
<value>com.mysql.jdbc.Driver</value>
<description>Driver class name for a JDBC metastore</description>
</property>
```

```
<property>
<name>javax.jdo.option.ConnectionUserName</name>
<value>hive</value>
<description>username to use against metastore database</description>
</property>

<property>
<name>javax.jdo.option.ConnectionPassword</name>
<value>hive</value>
<description>password to use against metastore database</description>
</property>

<property>
<name>hive.metastore.warehouse.dir</name>
<value>/user/hive/warehouse</value>
<description>location of default database for the warehouse</
description>
</property>
```

Modify the log file hive-log4j.properties.

```
#log4j.appender.EventCounter = org.apache.hadoop.metrics.jvm.
EventCounter
log4j.appender.EventCounter = org.apache.hadoop.log.metrics.
EventCounter
```

Set the environment variables.

```
~ sudo vi/etc/environment

PATH = "/usr/local/sbin:/usr/local/bin:/usr/sbin:/usr/bin:/sbin:/bin:/
usr/games:/usr/local/games:/home/cos/toolkit/ant184/bin:/home/cos/
toolkit/jdk16/bin:/home/cos/toolkit/maven3/bin:/home/cos/toolkit/
hadoop-1.0.3/bin:/home/cos/toolkit/hive-0.9.0/bin"

JAVA_HOME =/home/cos/toolkit/jdk16
ANT_HOME =/home/cos/toolkit/ant184
MAVEN_HOME =/home/cos/toolkit/maven3

HADOOP_HOME =/home/cos/toolkit/hadoop-1.0.3
HIVE_HOME =/home/cos/toolkit/hive-0.9.0

HADOOP_STREAMING =/home/conan/hadoop/hadoop-1.0.3/contrib/streaming/
hadoop-streaming-1.0.3.jar
CLASSPATH =/home/cos/toolkit/jdk16/lib/dt.jar:/home/cos/toolkit/
jdk16/lib/tools.jar
```

Create Hive directory on hdfs.

```
$HADOOP_HOME/bin/hadoop fs -mkidr/tmp
$HADOOP_HOME/bin/hadoop fs -mkidr/user/hive/warehouse
$HADOOP_HOME/bin/hadoop fs -chmod g+w/tmp
$HADOOP_HOME/bin/hadoop fs -chmod g+w/user/hive/warehouse
```

Create a database in MySQL.

```
create database hive_metadata;
grant all on hive_metadata.* to hive@'%' identified by 'hive';
grant all on hive_metadata.* to hive@localhost identified by 'hive';
ALTER DATABASE hive_metadata CHARACTER SET latin1;
```

Upload the jdbc library of MySQL to hive/lib manually.

```
~ ls/home/cos/toolkit/hive-0.9.0/lib
mysql-connector-java-5.1.22-bin.jar
```

Start the Hive server.

```
# Start metastore service.
~ bin/hive --service metastore &
Starting Hive Metastore Server

# Start HiveServer service.
~ bin/hive --service hiveserver &
Starting Hive Thrift Server

# Start Hive client.
~ bin/hive shell
Logging initialized using configuration in file:/root/hive-0.9.0/
conf/hive-log4j.properties
Hive history file =/tmp/root/hive_job_log_root_201211141845_1864939641.
txt

hive> show tables
OK
```

Now we've finished the installation of Hive on Linux Ubuntu®.

Appendix H: Installation of HBase

H.1 Environment Preparation of Ubuntu

HBase is distributed NoSQL database software based on Java and run on Hadoop. Because HBase doesn't provide a Windows® version, I'll just introduce the installation of HBase in Linux Ubuntu®.

HBase is run on Hadoop, so we need to install the Hadoop environment first. For the installation of Hadoop, please refer to Appendix F. Hbase doesn't provide an apt-get software source for installation, so we need to download the HBase package on the official website, http://www.apache.org/dyn/closer.cgi/hbase/.

System environment:

- Linux Ubuntu 12.04.2 LTS 64bit server
- Java JDK 1.6.0_45
- Hadoop 1.1.2

H.2 Installation of HBase

H.2.1 Download HBase

```
# Download through wget command.
~ wget http://www.gaidso.com/apache/hbase/stable/hbase-0.94.18.tar.gz

# Decompress HBase.
~ tar xvf hbase-0.94.18.tar.gz

# Move HBase directory to folder.
~ mv hbase-0.94.18//home/conan/hadoop/

# Enter the directory.
~ cd/home/conan/hadoop/hbase-0.94.18
```

H.2.2 Configure HBase

Modify the start-up file hbase-env.sh.

```
~ vi conf/hbase-env.sh

# Open the comment.
export JAVA_HOME =/home/conan/toolkit/jdk16
export HBASE_CLASSPATH =/home/conan/hadoop/hadoop-1.1.2/conf
export HBASE_MANAGES_ZK = true
```

Modify the configuration file hbase-site.xml.

```
~ vi conf/hbase-site.xml

<configuration>
  <property>
    <name>hbase.rootdir</name>
    <value>hdfs://master:9000/hbase</value>
  </property>

  <property>
    <name>hbase.cluster.distributed</name>
    <value>true</value>
  </property>

  <property>
     <name>dfs.replication</name>
     <value>1</value>
  </property>

  <property>
    <name>hbase.zookeeper.quorum</name>
    <value>master</value>
  </property>

  <property>
     <name>hbase.zookeeper.property.clientPort</name>
     <value>2181</value>
  </property>

  <property>
    <name>hbase.zookeeper.property.dataDir</name>
    <value>/home/conan/hadoop/hdata</value>
  </property>
</configuration>
```

Copy the configuration file and class library of the Hadoop environment.

```
~ cp ~/hadoop/hadoop-1.1.2/conf/hdfs-site.xml conf/
~ cp ~/hadoop/hadoop-1.1.2/hadoop-core-1.1.2.jar lib/
~ mkdir/home/conan/hadoop/hdata
```

H.2.3 Start Hadoop and the HBase Server

```
~/home/conan/hadoop/hadoop-1.1.2/bin/start-all.sh
~/home/conan/hadoop/hbase-0.94.18/bin/start-hbase.sh

# View HBase process.
~ jps
13838 TaskTracker
13541 JobTracker
15946 HMaster
16756 Jps
12851 NameNode
13450 SecondaryNameNode
13133 DataNode
15817 HQuorumPeer
16283 HRegionServer
```

H.2.4 Open the HBase Command Line Client to Access HBase

```
~ bin/hbase shell
HBase Shell; enter 'help<RETURN>' for list of supported commands.
Type "exit<RETURN>" to leave the HBase Shell
Version 0.94.18, r1577788, Sat Mar 15 04:46:47 UTC 2014
hbase(main):002:0> help # View the help information of HBase command
line.
```

H.3 Installation of Thrift

After we installed HBase, we also need to install Thrift because other languages need to use Thrift to call HBase. Thrift needs to be compiled locally, and the official website doesn't provide a binary package, so we need to download thrift-0.9.1, which is available at http://thrift.apache .org/download.

H.3.1 Download Thrift

There are two ways of downloading Thrift: download directly or download the source code through Git. The first choice is to download the source code distribution thrift-0.9.1.tar.gz directly.

```
~ wget http://apache.fayea.com/apache-mirror/thrift/0.9.1/thrift-
0.9.1.tar.gz
~ tar xvf thrift-0.9.1.tar.gz
~ mv thrift-0.9.1//home/conan/hadoop/
~ cd/home/conan/hadoop/
```

Note: All the errors below are caused by this distribution, so it is not recommended.

The second is downloading source code through Git.

```
~ git clone https://git-wip-us.apache.org/repos/asf/thrift.git
thrift-git
~ mv thrift-git//home/conan/hadoop/
~ cd/home/conan/hadoop/
```

To avoid all kinds of errors, we suggest that the second way be used.

H.3.2 Install Thrift through Git Source Code

Thrift needs to be compiled locally.
    Run installation commands.

```
# Enter thrift-git directory.
~ cd/home/conan/hadoop/thrift-git

# Copy 0.9.1 label to new branch thrift-0.9.1.
~ git checkout -b thrift-0.9.1 0.9.1

# Generate configuration script.
~./bootstrap.sh

# Generate configuration information.
~./configure

# Compile Thrift.
~ make

# Install Thrift.
~ sudo make install
```

Thrift is now installed through the Git source code version. Then view the Thrift version.

```
~ thrift -version
Thrift version 0.9.1
```

Next, start the Thrift server of HBase.

```
# Start Thrift Server of HBase.
~/home/conan/hadoop/hbase-0.94.18/bin/hbase-daemon.sh start thrift
starting thrift, logging to/home/conan/hadoop/hbase-0.94.18/bin/../
logs/hbase-conan-thrift-master.out

# Check system process.
~ jps
13838 TaskTracker
13541 JobTracker
15946 HMaster
32120 Jps
12851 NameNode
13450 SecondaryNameNode
13133 DataNode
32001 ThriftServer
15817 HQuorumPeer
16283 HRegionServer
```

Thrift is started, and now we can use various languages to access HBase through Thrift.

Bibliography

Couture-Beil, Alex (2014). "Package rjson Reference manual," available at http://cran.r-project.org/web/packages/rjson/rjson.pdf.

Lewis, Bart W. (2014). "Package rredis Reference manual," available at http://cran.r-project.org/web/packages/rredis/rredis.pdf.

Lewis, Bart W. (2014). "The rredis Package," available at http://cran.r-project.org/web/packages/rredis/vignettes/rredis.pdf.

Ooms, Jeroen (2014). "Package RMySQL Reference manual," available at http://cran.r-project.org/web/packages/RMySQL/RMySQL.pdf.

Owen, Sean, Robin Anil, Ted Dunning, Ellen Friedman (2014). "*Mahout in Action*," Beijing, Post & Telecom Press, pp. 91–114.

RevolutionAnalytics (2012). "RHadoop," available at https://github.com/RevolutionAnalytics/RHadoop/wiki.

RForge.net. "Examples for Rserve," available at http://rforge.net/Rserve/example.html.

Richet, Yann (2014). "Rsession: R sessions wrapping for Java," available at https://github.com/yannrichet/rsession.

Ryan, Jeffrey A. (2014). "Working with xts and quantmod Leveraging R with xts and quantmod for Quantitative Trading," available at http://www.rinfinance.com/RinFinance2009/presentations/xts_quantmod_workshop.pdf.

Ryan, Jeffrey A., Joshua M. Ulrich (2014). "Package xts Reference manual," available at http://cran.r-project.org/web/packages/xts/xts.pdf.

Ryan, Jeffrey A., Joshua M. Ulrich (2014). "xts: Extensible Time Series," available at http://cran.r-project.org/web/packages/xts/vignettes/xts.pdf.

Santini, Alberto. "RIO," available at https://github.com/albertosantini/node-rio.

Selivanov, Dmitriy (2014). "Package rmongodb Reference manual," available at http://cran.r-project.org/web/packages/rmongodb/rmongodb.pdf.

Shin, Bruce (2014). "Package RHive Reference manual," available at http://cran.r-project.org/web/packages/RHive/RHive.pdf.

Temple Lang, Duncan (2014). "Package RJSONIO Reference manual," available at http://cran.r-project.org/web/packages/RJSONIO/RJSONIO.pdf.

Tuszynski, Jarek (2014). "Package caTools Reference manual," available at http://cran.r-project.org/web/packages/caTools/caTools.pdf.

Urbanek, Simon (2008). "FastRWeb: Fast Interactive Web Framework for Data Mining Using R," available at http://urbanek.info/research/pub/urbanek-iasc08.pdf.

Urbanek, Simon (2014). "Package RCassandra Reference manual," available at http://cran.r-project.org/web/packages/RCassandra/RCassandra.pdf.

Urbanek, Simon (2014). "Package rJava Reference manual," available at http://cran.r-project.org/web/packages/rJava/rJava.pdf.

Urbanek, Simon (2014). "Package RSclient Reference manual," available at http://cran.r-project.org/web/packages/RSclient/RSclient.pdf.

Urbanek, Simon (2014). "Package Rserve Reference manual," Chapter 4.1, available at http://cran.r-project.org/web/packages/Rserve/Rserve.pdf.

Urbanek, Simon, Jeffrey Horner (2014). "Package Cairo Reference manual," available at http://cran.r -project.org/web/packages/Cairo/Cairo.pdf.

Urbanek, Simon, Jeffrey Horner (2014). "Package FastRWeb Reference manual," available at http://cran.r -project.org/web/packages/FastRWeb/FastRWeb.pdf.

Wickham, Hadley (2014). "Package memoise Reference manual," available at http://cran.r-project.org/web /packages/memoise/memoise.pdf.

Wickham, Hadley (2014). "Package profr Reference manual," available at http://cran.r-project.org/web /packages/profr/profr.pdf.

Wikipedia. "Base64," available at http://en.wikipedia.org/wiki/Base64.

Wikipedia. "ROC Curve," available at http://en.wikipedia.org/wiki/Receiver_operating_characteristic.

Xie, Yihui (2014). "Package formatR Reference manual," available at http://cran.r-project.org/web/packages /formatR/formatR.pdf.

Zeileis, Achim (2014). "Package fortunes Reference manual," available at http://cran.r-project.org/web /packages/fortunes/fortunes.pdf.

Zeileis, Achim (2014). "Package zoo Reference manual," available at http://cran.r-project.org/web/packages /zoo/zoo.pdf.

Zeileis, Achim, Gabor Grothendieck (2014). "zoo: An S3 Class and Methods for Indexed Totally Ordered Observations," available at http://cran.r-project.org/web/packages/zoo/vignettes/zoo.pdf.

Index

Page numbers in italics indicate figures.